Directed Energy System Performance Prediction

For a listing of recent titles in the *Artech House Electronic Warfare Library*, turn to the back of this book.

Directed Energy System Performance Prediction

Graham V. Weinberg

ARTECH HOUSE
BOSTON | LONDON
artechhouse.com

Library of Congress Cataloging-in-Publication Data
A catalog record for this book is available from the U.S. Library of Congress.

British Library Cataloguing in Publication Data
A catalogue record for this book is available from the British Library.

ISBN 13: 978-1-68569-027-4

Cover design by Joi Garron

ARTECH HOUSE
685 Canton Street
Norwood, MA 02062

10 9 8 7 6 5 4 3 2 1

Contents

Preface 11

Acknowledgments 15

1 Introduction 17

1.1 An Introduction to DE 17

1.2 A Historical Sketch of DE Systems 20

1.2.1 High Power Radio Frequency 20

1.2.2 High Energy Lasers 22

1.3 A Selection of Systems 25

1.3.1 Stryker-Mounted Laser 25

1.3.2 U.S. Navy's Laser Weapon System 26

1.3.3 Boeing YAL-1 Airborne Laser Testbed 28

1.3.4 Epirus Leonidas High-Power Microwave Systems 29

1.3.5 Air Force Research Lab's Tactical High-Power Operational Responder 30

1.3.6 Other DE Systems 31

1.4 Purpose and Scope 34

References 35

2 Some Principles of Mathematics and Physics 39

2.1 Probability and Statistics Basics 40

2.1.1 Fundamentals 40

2.1.2 Distributions and Properties 42

2.1.3 Statistical Conditioning 44

2.2 Principles of Stochastic Processes 45

2.2.1 Fundamental Processes 45

2.2.2 Queueing Theory 47

2.3 Physics Preliminaries 50

2.3.1 Wavelengths and Frequency 51

2.3.2 Propagation of Electromagnetic Energy 52

2.3.3 Signals and Fourier Analysis 55

References 60

3 Fundamentals of HPRF Performance Modeling 63

3.1 An Overview of HPRF DEWs 63

3.2 Electronic Vulnerability Levels 66

3.3 HPRF Power Density Function 67

3.4 Narrowband and Wideband Effector Performance Modeling 73

3.4.1 Example: A Damped Sinusoidal Signal 76

3.4.2 Power Density for Wideband and Ultrawideband Signals 80

3.5 Concluding Remarks 82

References 82

4 HPRF Performance Prediction 85

4.1 One Threat 86

4.1.1 Narrowband Case 86

4.1.2 Wideband and Ultrawideband Case 89

4.2 Multiple Threats 92

4.2.1 Threats Arrive Simultaneously 92

4.2.2 Arrivals Through a Renewal Process 96

4.2.3 Threats Arrive Linearly 101

4.3 Concluding Remarks 107

References 108

5 The HEL Irradiance Function 109

5.1 Lasers: An Overview 110

5.2 Laser Damage Thresholds 112

5.3 Gaussian Beam Profiles 114

5.4 Irradiance Functions 118

5.5 Irradiance Function Examples 121

5.6 Some Final Comments on Irradiance Functions 130

References 133

6 HEL Performance Prediction 137

6.1 Models for Thresholds 138

6.2 Single Target and Single Effector 141

6.3 Multiple Targets: Queueing Theory Approach 146

6.4 Multiple Sources on a Single Threat 155

6.5 Number of Effectors to Achieve Minimum Performance 158

6.6 Concluding Remarks 161
References 162

7 Future Research Directions 163

7.1 HPRF DEW Considerations 163
7.2 Modeling Synchronized DEW Systems 164
7.3 Active Protection Systems 165
7.4 Adaptive Optics 165
7.5 Validation of Performance Models 166
7.6 Development of Performance Prediction for Acoustic Systems 166
References 166

A Emerging Threat Exemplars 169

A.1 Airborne Threats 169
A.2 Ground-Based Threats 171
A.3 Sea-Based Threats 172
References 175

B Irradiance Function for Adaptive Optics 177

References 181

C Distribution of Delays 183

Reference 186

List of Acronyms 187

List of Symbols 189

List of Units 191
About the Author 193
Index 195

Preface

From a military utility perspective, directed energy (DE) has been identified as a tool of significant importance for the current and future fighting forces worldwide, especially given the evolving modern combat environment where asymmetrical warfare is becoming prevalent. This has been made clear since the 2022 Russian invasion of Ukraine, where applications of new technology, such as uncrewed (formerly unmanned) aerial systems, has been witnessed in real time. The applications of DE are somewhat specialized, and are by no means a replacement for the tools of conventional warfare. However, current and emerging threats appearing in asymmetrical warfare such as uncrewed systems deployed across the three domains of air, sea, and land can be neutralized through DE weapons (DEWs).

The problem posed by uncrewed aerial vehicles (UAVs) is significant not only from a military perspective but also has ramifications for the security of civilians. As has been demonstrated in recent years, it is possible to control a large number of UAVs through a single laptop, as in displays provided by Intel during the Tokyo 2020 Olympic Games opening ceremony. In this demonstration Intel launched 1,824 UAVs to provide a spectacular night-sky light show performance. While this application of swarms of UAVs is aesthetically pleasing and benign, it nonetheless raises a number of concerns for both law enforcement and the military forces.

From a civilian perspective, UAVs may be used for surveillance purposes and so there is the potential for abuse by those of a criminal orientation. UAVs have the potential to disrupt airport traffic, as the 2018 Gatwick airport incident demonstrated adequately. The major concern is that UAVs may function as an improvised explosive device and so can be used for terrorist activities. Additionally, a swarm of nonweaponized UAVs may be used to disrupt or damage aircraft through collisions.

In the military context, the problems caused by UAVs are similar. A military base may be compromised by UAVs, causing damage to key assets or providing tactical information to an enemy force. From the point of view of a team of dismounted combat soldiers in the battlefield, UAVs may be providing the enemy with intelligence, or their presence may limit their freedom of movement. Swarms of UAVs may also be used to disrupt navy vessels for similar reasons, and especially in the case of an aircraft carrier.

These examples outline the issues with UAVs; similar problems arise for uncrewed systems designed to operate as semiautonomous robots as well as uncrewed sea vessels.

DE systems have demonstrated their capability in defeating threats such as those outlined above. In particular, high energy laser (HEL) DEWs have shown their effectiveness in neutralizing swarms of UAVs. Additionally, a HEL DEW may be used for precision strikes against either moving targets or infrastructure. High power radio frequency (HPRF) DEWs are also suitable for the problem of neutralizing swarms of UAVs, and also have utility in the disruption of communication systems. Both these DEWs have deployment issues but it is possible to specify operational situations where either DE system may be utilized successfully.

Although there are a number of useful references on both these DE systems and their development, what is needed is a guide on how their performance may be assessed. Hence this book is concerned fundamentally with the development of a performance prediction capability for DE systems. This is achieved by modeling mathematically DE performance from the point of view of measures of the average power level incident on a target. When a DE system transfers energy to a target it will cause it to malfunction by either a thermal effect or by causing its internal electronics to fail. The underlying premise adopted in this book is that once the average power level incident on a target exceeds the target's inherent vulnerability level, it may be considered neutralized.

From the DE perspective the arrival of threats may not be deterministic, and so their arrivals within the DEW's theater of operation is assumed to be sensed through an integrated detection and tracking process. The times of detection possess a duality with the threat arrival times, and so one may proceed by focusing on arrivals through a stochastic process. This principle will be utilized in both the performance prediction of HEL and HPRF DEWs.

It is important to emphasize that the focus in this book is on the development of a mathematical modeling framework for DEW performance prediction. Hence extensive discussions on the construction of such systems, and their integration into appropriate platforms, is out of scope.

For completeness, the book also includes a brief historical overview of DEWs and discussion of some exemplars. In addition to this, a survey of the relevant tools of mathematics and physics is provided, together with a series of pertinent references to which the reader is referred for further information.

For the reader's reference, Appendix A includes a discussion of exemplars of uncrewed systems, since these are viewed as a primary target suitably neutralized by DEW systems.

Organizationally, the book examines the performance prediction of HPRF and HEL DE systems separately; each has a chapter devoted to the construction of the function measuring the power incident on a target and then a chapter examining its application for performance prediction.

This book has been written as an introduction to performance prediction for DE systems; as will become apparent, there is sufficient scope to expand the developments outlined in the following chapters. To function as a catalyst for this, the final chapter outlines explicitly a number of potential future research directions that interested researchers are encouraged to pursue.

Acknowledgments

The support of the Australian Defence Science and Technology Group is acknowledged, and in particular those within this organization who have encouraged me and shown interest in this book writing project. Specifically, I would like to thank Fred Bowden and Nick Kempt for the opportunities they have provided me with since 2018. Discussions with Sean Braidwood on the formulation of the high power radio frequency power density function formed the basis for some aspects of this work. The encouragement and interest of Andrew Malouf and Mitchell Kracman has also been appreciated.

A large number of publicly-available images have been used throughout this book to illustrate exemplars of directed energy systems. In all cases approval from the copyright owner has been sought. In some cases the relevant point of contact kindly provided higher-resolution images. Hence, I would like to thank UVision, Epirus, the U.S. Army, Navy, and Airforce for their contributions.

Finally, I need to acknowledge the person who acted as a catalyst for this consolidation of research into this monograph: Milla Visotcky, who also was instrumental in the decision to write my first book *Radar Detection Theory of Sliding Window Processes*.

CHAPTER

1

Contents

1.1 An Introduction to DE

1.2 A Historical Sketch of DE Systems

1.3 A Selection of Systems

1.4 Purpose and Scope

References

Introduction

This chapter sets the context for the analysis to follow by providing a general introduction to directed energy (DE) and its application to the scenarios of interest. The discussion here will be technical in part but has been written from a generalist perspective. Subsequent chapters will examine DE in terms of mathematical and physical models for performance prediction. Included in this chapter will be brief historical overviews of the development of the two main DE systems of interest, as well as a synopsis of exemplars of systems that were either developed as prototypes or demonstrators, as well as some that have evolved into operational use.

1.1 An Introduction to DE

DE refers to the focusing of electromagnetic energy into a form so that a desired outcome can be achieved. When the term weapon is appended to DE, it refers to such focusing of energy in order to achieve a disruptive effect on a target. Consequently, the term directed energy weapon (DEW) will be employed throughout. The usage of the expression DE system in this book has been to emphasize the fact that DEWs are not the only permissible applications of DE. Although the primary focus in this book is on the application of DE for disruptive effects, other emerging applications of DE include remote powering of uncrewed (historically unmanned) aerial vehicles (UAVs) and the charging of mobile phones [1, 2].

In this context of DEWs, appropriate targets are threats such as UAVs, loitering munitions, ground-based semiautonomous robotic systems, and missiles. Appendix A provides some examples of these threats for the reader's interest. DEWs can also be used for long-range precision strikes against enemy military assets as well as for disruption of communications to reduce an enemy's command and control. In addition to this it has military utility for the long-range disruption of sensors and intercontinental missiles.

Although DE has been in existence for a considerable period, its emerging importance in asymmetrical warfare has stimulated western countries to consider its utility more carefully and systematically [3]. This has resulted in the United States (U.S.) producing a strategy for the next 40 years of DE system development as well as concepts of use and employment [4].

Since DEWs are transmitting electromagnetic energy they operate at speeds close to that of light in a vacuum; the reduction in speed can be due to attenuation in the environment in which the energy is transmitted. Hence they also have a significantly reduced visible as well as noise signature, and can operate with what is commonly referred to as a "bottomless magazine." However, this term needs to be somewhat qualified: They can operate continuously in the presence of a constant power supply and also conditionally on no system disruptions, such as malfunctions.

The two main types of DEWs under consideration in this book are high power radio frequency (HPRF) and high energy laser (HEL) systems. The main difference between these two are their operational wavelengths, which consequently have an impact on their functional characteristics. HEL DEWs function in very small wavelengths, and so a beam of energy can remain coherent over long distances, provided the energy is not sufficiently attenuated, distorted, or subject to jittering. In contrast to this an HPRF DEW operates in much larger wavelengths, and has the disadvantage of possessing a beam pattern that can diverge rapidly with range. Hence a HEL DEW is often considered an appropriate tool for precision strikes against threats at long distances, while a HPRF DEW can be employed for regional defense against localized airborne threats.

There is also a difference in the way in which these DEWs interact with a target to achieve a disruptive effect. A HEL DEW focuses energy in order to achieve a thermal disruption, although it may also be used to produce temporary damage to a target, especially if the threat has optical vulnerabilities. By contrast, an HPRF DEW applies pulsed energy to induce surface currents on the target, consequently causing system coupling. While a HEL DEW can focus its beam on any nonreflective exposed surface on the threat, a HPRF DEW disruption may either be achieved through transmitting waveforms at a frequency matched to an artifact on the target, such as an antenna, or coupling may occur if waveforms enter an exposed aperture. Currents flowing on the surface on a target may also induce currents inside the threat and consequently damage electronic systems. These two ways in which coupling may occur are referred to as front door and back door coupling, and will be discussed further in Chapter 3.

One important observation concerning HPRF DEWs is that it is also extremely difficult to provide sufficient shielding against them, since any airborne system will possess unshielded antennas or sensors, with an exception being a "dumb" airborne missile that has no guidance or control. This shielding factor also introduces a complexity from an operational perspective. In simple terms since an HPRF DEW can cause intentional interference in electronic systems it has the potential to disrupt radar systems coupled with it for automatic operation. Hence it is necessary to consider this issue in the deployment of such DEWs. Consequently this may also limit the military utility of HPRF DEWs. To clarify this, while a stand-alone HPRF DEW system in a land-based setting may be fit for purpose, one deployed on a naval vessel may be more problematic. Some of these issues will be examined further in Chapters 3 and 4 concerned with HPRF DEWs.

A useful classification of DEWs is in terms of the effects that they cause in a threat. Hard kill refers to the case where a DEW is applying energy to cause a permanent disruption, or even destroy the threat directly. Clearly HEL DEWs fall into the hard kill category, while HPRF DEWs can also result in a hard kill. By contrast, a soft kill refers to the situation where a target is only disrupted temporarily. Both HEL and HPRF DEWs can also operate in such a manner. In fact, an HPRF DEW can cause disruptive effects akin to electronic warfare, since they can be used to disrupt communication and information systems. However, DE systems aim to degrade or destroy, whereas an electronic warfare system may be used to deceive, as in the case of sensor jamming to mask the presence of a combatant.

HPRF DEWs have a number of applications, some of which can also be performed by HEL DEWs. A large scale HPRF DEW can be utilized for regional defense against airborne threats. In order to generate sufficient energy to provide such defense, a large power supply is required as well as a sufficiently large directive antenna to deliver the energy towards the target. However, HPRF DEWs can be reduced in size by limiting their power and the manner in which their waveforms are generated. This means that they can also be more readily deployed for applications such as short-range communication and vehicle disruptions.

It is of interest to provide a survey of the historical roots of DE, and so the next section provides brief accounts focused on the two DE systems of interest in this book.

1.2 A Historical Sketch of DE Systems

When discussing the beginnings of DE, reference is often made to the legendary story where the famous Greek mathematician Archimedes destroyed the Roman fleet in the siege of Syracuse in Sicily during 213 B.C., by focusing the Sun's rays through large mirrors [5]. However this story has now been somewhat discounted through scientific studies [6]. Nonetheless it provides an interesting perspective on the history of DE since it suggests that the focusing of energy for regional defense may have been of consideration in the ancient world. Science fiction authors have also with prescience envisioned DE systems; a famous example is the Martian heat ray in H. G. Wells' *War of the Worlds* first published in 1898 [7]. An interesting discussion of science fiction conceptions of DE-like systems during the 1920–1930s can be found in [8], which discusses the interesting duality between science fiction conception of focused disruptive energy and the emergence of such technology.

Since the focus in this book is on HPRF and HEL DEWs, the following historical accounts are focused on their development only, both of which have their well-documented origins in twentieth-century scientific developments.

1.2.1 High Power Radio Frequency

The origins of the study of HPRF effects can be found in the nuclear tests in the 1940–1960s, and in particular it was during the 1945 Trinity Event in New Mexico that the effects of an electromagnetic pulse (EMP) were first observed. This effect was actually anticipated by the theoretical predictions of the famous Italian physicist Enrico Fermi, who had electronic equipment shielded in expectation of electromagnetic effects from the explosion. An EMP was found to result from an effect caused by gamma rays emitted during a nuclear explosion. Specifically, they produce a current of Compton recoil electrons resulting in further ionization and consequently resulting in air becoming conductive. This then results in the generation of an electromagnetic field [9]. The gamma rays are produced in a pulse with an extremely fast rise time in the order of a few nanoseconds. The British nuclear tests of the 1950s also reported the effects of the EMP on electronics, referred to as radioflash. A more significant test was in 1962, where the U.S. detonated a megaton nuclear weapon 400 km above the Pacific Ocean, whose resultant EMP damaged electronic and electrical systems in Hawaii, which is over 1,600 km away from the test site. Observations of the EMP, and its effects on electronics, resulted in much interest in its potential as a defensive weapon or countermeasure to missile strikes in the 1960s. Hence, high altitude EMP was identified as of strategic importance during the Cold War, with both the United States and the Union of Soviet Socialist Republics (USSR) pursuing research interests in it [10].

The study of EMP was complicated by the fact that inadequate sensors were available to measure the resultant electromagnetic fields. Underground nuclear tests were found to facilitate the study of EMP effects. However, it was discovered that EMP could be examined through the effects of natural lightning strikes, since for surface burst EMP the currents induced in buried cables are similar to those induced by lightning. Hence an understanding of the lightning source field effects was expected to facilitate the study of EMP [11].

The analysis of the impact of EMP fields and lightning strike–induced fields on aircraft helped identify electromagnetic characteristics likely to induce effects on electronic equipment [12]. It was found that for frequencies below 1 MHz, direct lightning strike–induced fields were dominant in producing effects, while for frequencies above 10 MHz the field induced by a high altitude EMP dominated. These results yielded useful information on the relative field strengths and frequencies required to disrupt electronics through induced electromagnetic fields.

Given these results, focus shifted to conventionally produced high-power electromagnetic fields through the study of source generation and antennas, although there appears to be a resurgence of interest in EMP in recent times [13, 14]. High power microwave tubes, and in particular relativistic magnetrons, became of interest, which work well in the 1-GHz frequency range, and are able to generate peak powers in the order of 1 GW with a pulse length of 100 nanoseconds. As observed in [11], frequencies in this range are the best for disrupting electronics through coupling. Consequently, during the 1970s there was significant work undertaken in both the United States and Soviet Union on the construction of effective narrowband HPRF effectors to replicate the effects of EMP on electronic equipment.

An interesting historical account of the DE work by the U.S. Special Applications branch of the Naval Weapons Laboratory at Dahlgren, Virginia is [15], where it is pointed out that during the 1960s extensive work was performed in developing ways to generate high-power oscillating electric fields that could be exploited for tactical purposes. During the 1970s Hertzian oscillators were thus examined, resulting in the application of Marx generators for voltage multiplication. Interest in pulse-power systems gained momentum in around 1976, since it was realized the reduction in system size really depended on the reduction of the mechanism delivering the pulsed power. Consequently, more effort was invested in the required power delivery technology.

Pulse power components enabled energy to be stored over long periods and released very rapidly, resulting in massive peak power levels. As a result of this, the development of HPRF DE systems focused on compact design of pulsed power effectors, and during the 1980s and 1990s there was considerable advances reported in the literature.

With the emergence of compact HPRF effectors a new threat to society evolved, which became knows as intentional electromagnetic interference (IEMI) [16]. This refers to the application of HPRF DE systems to the disruption of civilian infrastructure in criminal and terrorist acts [17]. Thus a whole field grew out of IEMI, examining its impacts and ways in which it can be countered [18, 19].

1.2.2 High Energy Lasers

An excellent and accessible historical account of the evolution of lasers and quantum physics is the book [20], while [21] provides a useful timeline of the developments in laser theory and technology. The genesis of lasers resulted from the understanding that light is fundamentally a form of radiation, and this grew from the development of quantum mechanics during the early decades of the twentieth century. This was the consequence of the pioneering work by Albert Einstein in 1917, where the process of stimulated emission was introduced in a paper on the quantum theory of radiation [22]. However, quantum theory began with the discovery attributable to Max Planck that energy is radiated and absorbed in discrete quanta [23]. This result, published in 1900, was encapsulated in his equation showing that electromagnetic energy is a constant multiple of its frequency. This enabled Planck to accurately explain phenomena observed in black-body radiation [24].

Two Soviet physicists, Alexander Prochorov and Nikolai Basov, published a paper, written in Russian in October 1954, describing the theory that Einstein's stimulated emission work could be used to amplify microwave energy, through a process described as microwave amplification through the stimulated emission of radiation (maser) [25]. These Soviet scientists presented their work at a conference in 1955, and an American physicist Charles H. Townes, who was also at the conference, was surprised by their presentation because he had already built an ammonia maser as described by Prochorov.

The background to this was that Townes worked on microwave radar under a Bell Laboratory radar project during the Second World War. His objective was to produce a radar operating in the higher GHz frequency band and he demonstrated that the gas ammonia was easily excited by light of around 24 GHz. His objective was to produce a more accurate radar, but it turned out that radars operating at this frequency have extremely poor performance.

Townes moved to Columbia University and in 1954, working jointly with Herbert J. Zeiger and James P. Gordon, demonstrated a maser radiating at a wavelength of over 1 cm and generating 10 mW of power. Following this, in 1958 Townes and Arthur L. Schawlow published a seminal paper in *Physical Review* explaining how masers could be made to operate in the optical and infrared spectrum [26]. Simultaneously, the two Soviet scientists Prochorov and Basov were also exploring this extension of masers.

In 1957, a Columbia University graduate student named Gordon Gould proposed that stimulated emission could be used to amplify light, and coined the laser acronym. Subsequently, he left the university to join the private research company Technical Research Group (TRG). During April 1959, Gould and TRG applied for a patent stemming from Gould's laser ideas. TRG had contracts with the U.S. Department of Defense and it was with Gould's joining TRG that military interest in laser technology began.

TRG suggested that Gould's laser project could be funded though the U.S. Department of Defense. This proved to be a good idea because the United States had set up the Advanced Research Project Agency (ARPA) in view of Soviet progress in technology during the early stages of the Cold War. Hence, the day after the publication of Townes and Schawlow's paper, TRG submitted a proposal to the Pentagon, requesting $300,000 in funding. Instead ARPA funded them one million dollars, not only funding the laser development but also funding TRG for a range of possible military applications for the developed laser.

The first working laser was developed by Theodore H. Maiman, a physicist with Hughes Research Laboratories, in May 1960. He published a short paper on this in *Nature* three months later, describing its construction from a cylinder of synthetic ruby [27]. Following this there was a rapid development of laser prototypes, exploring different mechanisms to produce a beam [28]. For example, in November 1960 a uranium laser was produced by Peter P. Sorkin and Mirek J. Stevenson of IBM's Thomas J. Watson Research Center. In December 1960, Bell Laboratories produced a helium-neon laser, generating a continuous beam of light at a wavelength of 1.15 μm. In October 1961, the American Optical Company produced the first neodymium glass laser, while in June 1962, Bell Laboratories manufactured the first yttrium aluminum garnet laser.

One of the earliest reported military applications of lasers was in the Falkland's War during 1982, where the United Kingdom deployed laser weapons to dazzle Argentinian pilots [29].

There were a lot of controversies with the development of laser systems, in contrast to the genesis of HPRF DE, mainly attributable to ambiguity as to who developed which system first, resulting in a long-term patent war; the reader can consult [20] for an interesting account of this, as well as descriptions of the nonmilitary applications of lasers.

This completes the brief historical sketch of the evolution of HPRF and HEL DE. To complement this, the next section will provide some examples of the progress made in developing DE systems for applications of interest in this book.

1.3 A Selection of Systems

It is of relevance to include a survey of some of the DE systems that have been deployed as prototypes or operational systems, since this will provide an indication of the scope and potential utility of DE systems. This survey is not comprehensive and includes exemplars where relevant information is publicly available. It is important to note that, not surprisingly, explicit details of these systems are not readily available. Regardless of this, it is possible to speculate on operational characteristics of such systems.

1.3.1 Stryker-Mounted Laser

The Stryker is an eight-wheeled armored fighting vehicle (AFV) produced by Canada's General Dynamics Land Systems for the U.S. Army. Raytheon, in collaboration with the system integration company Kord, has successfully deployed a 50-kW HEL into a Stryker combat vehicle, as illustrated in Figure 1.1. The integrated system includes a specialized radar target acquisition system, a beam controlling system, and targeting sensors. The HEL provides 360°-perimeter defense against threats. It has been reported that this

Figure 1.1 The Raytheon 50-kW HEL integrated into a Stryker AFV. (Source: U.S. Army.)

system has been effective not only against a diverse range of UAVs but also in countering 60-mm mortar rounds [30]. This impressive exemplar shows that many of the size, weight, and power (SWaP) issues have been resolved in terms of integrating a sufficiently powerful HEL into an AFV.

The German company Rheinmetall has also developed a suite of HEL systems, with very impressive performance capabilities, and some of which have been integrated into AFVs. Relative to power levels, they provide useful details on the success of various HELs in a number of contexts [31]. Their Skyshield HEL effector, which is significantly large suggesting that it is for either fixed land-based defense or naval applications, is reported to have an effective range of up to 3 km, with a power level of 30 kW.

A series of vehicles were fitted with scaled down HELs and Rheinmetall reports on the results of tests, which are useful indicators of expected performance relative to power levels. A 5-kW HEL was integrated into an M113 armored personnel carrier, and shown to be successful in terms of detecting and neutralizing battlefield ordnance, such as land mines, at a safe distance. As a second example, a 20-kW HEL was integrated into a GTK Boxer AFV, and an interesting experiment showed that the system could be used to neutralize a heavy machine gun mounted on a pickup truck. This was done by applying the laser directly to the machine gun to cause it to malfunction without any damage to the gunner (in this case a mannequin). This system was also shown to be suitable for counter UAV applications when integrated with their targeting radar.

1.3.2 U.S. Navy's Laser Weapon System

The next DE system considered is one that has been developed for maritime operations. This is the U.S. Navy's Laser Weapon System (LWS), which was originally developed for deployment on the USS Ponce, an Austin-class amphibious transport dock. The system itself was built by Kratos Defense and Security Solutions. The integrated system was tested in 2014 and subsequently approved for defensive operations [32]. Figure 1.2 shows the laser aperture on the ship.

Its power has been estimated to be in the 15- to 50-kW range for counter UAV, defending against small aircraft and boats. The LWS is an AN/SEQ-3 laser designed primarily for low-end asymmetrical threats. It is a scalable system so that it can either be used for hard-kill (thermal effects) or soft-kill (disabling sensors). Its design had to take into consideration the complexity

Figure 1.2 The U.S. Navy's laser weapon system deployed on the USS Ponce. (Source: U.S. Navy.[1])

of operating in a maritime environment. In addition to this, a chemical-based laser was not considered safe for deployment on a ship and so necessitating a solid-state fiber laser system. It was actually deployed in the Persian Gulf with the U.S. Navy's 5th Fleet in 2014 where it demonstrated that it was able to function in the maritime environment without performance complexities.

After the USS Ponce was decommissioned in 2017, the LWS was integrated into the USS Portland, which is a San Antonio-class amphibious transport dock. It is reported that the deployed system had 150 kW of power to defeat a range of threats. There have been several reports of its success in terms of military utility in defeating a range of targets [33].

There are many more examples of HELs developed for naval deployment, and [34] provides a useful overview. As an example, Lockheed-Martin has produced what is known as the High-Energy Laser and Integrated

1. Use of released U.S. Navy imagery does not constitute product or organizational endorsement of any kind by the U.S. Navy.

Optical-Dazzler with Surveillance (HELIOS) system. This has been designed to provide naval platforms not only with a powerful HEL but also includes long-range intelligence, surveillance, and reconnaissance (ISR) capabilities integrated into the system. In addition to this, it can also function in soft-kill mode, targeting sensors on UAVs for example.

1.3.3 Boeing YAL-1 Airborne Laser Testbed

The third exemplar considered is one that was deployed in an airborne platform. The Boeing YAL-1 was an airborne laser testbed, with a HEL integrated into a modified Boeing 747-400F, designed for antiballistic applications [35]. It employed a MW-class chemical oxygen iodine laser, developed by Northrop Grumman, while the nose-turret and fire control system was produced my Lockheed Martin. Figure 1.3 shows the aircraft with the HEL mounted in its nose-turret. The system also included two kW-class lasers for target detection and tracking purposes.

This system was tested in March 2007 where it defeated a NC-135E test aircraft in flight. This demonstrated the fact that the system could compensate for the effects of atmospheric distortion [36]. The YAL-1 was decommissioned in September 2014 but it is reported that Lockheed Martin is developing tactical airborne laser pods for fighter jets [37].

Figure 1.3 The Boeing YAL-1 Airborne Laser Testbed. (Source: U.S. Missile Defense Agency.)

1.3.4 Epirus Leonidas High-Power Microwave Systems

The U.S. company Epirus has introduced a SWaP-reduced series of portable HPRF effectors, based upon new solid-state technology [38]. In particular their class of Leonidas High-Power Microwave Systems have shown much usefulness [39]. Their systems utilize a digitally beamformed antenna allowing optimization of power on a threat while minimizing the effects of unintentional interference. Figure 1.4 is a picture of the system deployed on a military-grade trailer. The system can be integrated into an AFV as illustrated in Figure 1.5, enabling a more portable HPRF DE defensive solution.

Epirus has also developed a lightweight compact HPRF system that can be integrated into a sufficiently sized UAV such as a quadcopter, as illustrated in Figure 1.6. Known as the Leonidas Pod this can be coordinated with the ground-based exemplar of Figure 1.4. Its main utility is for air-to-air engagements, such as in neutralizing swarms of UAVs, and has been reported to perform well in trials [40].

Figure 1.4 The Leonidas high-power microwave system. (Source: Epirus Inc.)

Figure 1.5 Deployment of the Leonidas system on AFVs. (Source: Epirus Inc.)

The advantage of the Leonidas Pod is that it is a multishot recoverable effector. It is interesting to compare it with what is known as the Counter-Electronics High Powered Advanced Missile Project (CHAMP), which was a development jointly between Boeing Phantom Works and the U.S. Air Force Research Laboratory Directed Energy Directorate [41]. CHAMP is a missile equipped with the facility to deliver an intense EMP during its trajectory, in order to disrupt electronics, and then ultimately it will self-destruct. It has been reported that in a 2012 experiment in the Utah desert it successfully disabled a series of electronic systems, while inadvertently disabling surveillance cameras and other sensors that were set up to monitor its progress [42].

1.3.5 Air Force Research Lab's Tactical High-Power Operational Responder

Built by the Air Force Research Lab (AFRL) Directed Energy Directorate at the Kirtland Air Force Base, the Tactical High-Power Operational Responder (THOR) is a large scale HPRF DEW. Its primary purpose is to provide regional

Figure 1.6 Leonidas Pod deployed on a quadcopter. (Source: Epirus Inc.)

defense against airborne threats, and in particular swarms of UAVs. As Figure 1.7 illustrates, this system is deployed in a 20-foot shipping container, and is transported by a C-130 transport aircraft. The system characteristics are not known publicly but given the size of the system it is expected that it would provide very good protection against airborne threats [43].

On the same scale as THOR is the Raytheon Phaser [44], which is also mounted in a shipping container and reported to have excellent performance against UAVs. A key difference is in the antenna design, since the Phaser has a reflector with a steering mirror, while THOR has a directional parabolic dish antenna.

1.3.6 Other DE Systems

There are other DE systems that may be of interest to the reader but are considered out of scope for the current study. Nonetheless it is useful to provide a concise overview of some of the more interesting examples.

Figure 1.7 The THOR HPRF system, developed by the U.S. AFRL. (Source: AFRL Directed Energy Directorate.)

During the 1980s there was much interest in the utility of particle beams for defense against intercontinental missiles, and the U.S. Strategic Defense Initiative was concerned with the development of a neutral particle beam effector deployed in space [45]. Particle beams apply a high energy beam of subatomic particles to cause target damage through interactions of particles at the molecular level. The main issue is the generation of the beam through particle accelerators, a problem when SWaP is of importance, as well as maintaining beam stability over vast distances [46].

The Active Denial System (ADS), and its later genesis known as the Silent Guardian^{TM} Protection System, was developed for the U.S. Army by Raytheon as a nonlethal, antipersonnel DE source [47]. This system operates at a wavelength of 3.2 mm and functions in a similar way to a microwave oven. Hence it is designed to cause heating on the surface of the target's skin, although at its operational wavelength the heating is only superficial. The effect is equivalent to what one experiences when opening an oven's door when it is fully heated. Although the system was deployed to Afghanistan it was not apparently used in operations [48]. There is speculation that this system may have unexpected results in certain operational environments, since microwave energy can be attenuated by atmospheric conditions.

The final example considered is known as the Long-Range Acoustic Device (LRAD), which has also been developed as a nonlethal crowd control mechanism [49]. However, it has found application in military contexts, especially for naval ship security [50]. It is reported that, as a deterrent weapon, it can emit sound waves at a distance exceeding 8 km at a peak level of 162 dB. This sound level is sufficient to cause severe pain in the recipient, and prolonged exposure may cause permanent hearing impediments [51]. As an aside it is worth noting that the approach introduced in this book may also be applied to quantify the performance of acoustic directed energy systems, such as the LRAD. This will be considered further in Chapter 7.

At this juncture it is worth discussing some historical examples of DE that have been the subject of much controversy. The first is known as the *Moscow Signal*, where during the early 1950s to mid-1970s U.S. sources detected a low-power RF signal being directed toward the U.S. embassy in Moscow [52]. The signal was identified as originating from an apartment building 100m away from the embassy, and the signal itself varied in the 2.5- to 4.0-GHz range with a field strength of roughly 0.05 watts per square meter. Although this field strength is weak there were later concerns that this microwave energy exposure had caused medical issues for staff in the embassy [53, 54].

A second notable example is known as the *Havana syndrome*, which is the term used to describe unexplained medical conditions experienced by American and Canadian embassy staff posted abroad [55]. This was first observed by staff in the U.S. embassy in Havana, Cuba in 2016. Staff reported hearing grating noises in specific directions, followed by symptoms such as hearing and memory loss as well as nausea. A number of studies have identified several possible causes, including sonic and microwave DEWs [56].

There are also more recent examples of applications of DE in questionable circumstances. Laser pointers have gained a certain level of notoriety due to their use in antisocial contexts. An example of this is the protests in Hong Kong during 2019 against increased Chinese influence and control. It is reported that protestors used laser pointers against local law enforcement personnel, focusing them on individual's eyes in order to cause temporary blindness [57].

1.4 Purpose and Scope

The primary focus of this book is to provide a systematic study of the performance prediction of DE systems, introducing useful tools for this purpose. DE systems are viewed as suitable for certain military purposes, with the target of the system being exclusively an uncrewed system. Hence it is important to emphasize that the applications of DE discussed in Section 1.3.6 to cause injury to humans is considered an unacceptable use of the technology.

There are many available references on HPRF and HEL DEWs, as well as a number of useful sources on DEWs in general. Suitable introductions to the concepts of DE include [3] and [58], while [59–62] are useful introductions to HPRF DE. From the HEL DE perspective, [63–66] are relevant sources. In view of these information sources it is therefore important to delineate the scope and focus of this book. As the reader will discover, this book is not focused on the physical design or technical specifications of DEWs, nor is it concerned with the integration of these systems into suitable platforms. Instead it is focused on the development of a framework in which the performance prediction of a DE system can be undertaken. In the context of DEW this is to facilitate an assessment of its operational effectiveness. This will be done exclusively through analysis of the HPRF power density on target, and via the irradiance function for HEL DEWs. Both these functions model the DEW power on the target per unit area, once one accounts for environmental attenuating factors.

The statistical modeling of the performance of DE requires the identification of appropriate stochastic features of the system. When an HPRF DEW applies energy towards a target, there are a number of stochastic factors that come into consideration. In the first case, the arrival time of threats into the DEW's operational beam will be random, and so it is necessary to model this appropriately. Secondly, once a threat is in the DEW's beam, the surface area on the threat that is likely to induce couplings is known as the coupling cross section, and being stochastic in nature, must be modeled appropriately.

The HEL DEW case is different with the exception of arrival times of targets into the DEW's energy beam still being random. As will be observed, the HEL irradiance on a target is deterministic; the stochastic elements will be the time it takes to cause an effect and the target threshold vulnerability level.

Hence the book aims to demonstrate how these quantities can be expressed statistically and modeled through stochastically-driven processes.

References

[1] https://news.usni.org/2019/05/08/navy-studying-how-lasers-could-power-unmanned-vehicles, 2019.

[2] https://www.washington.edu/news/2018/02/20/using-a-laser-to-wirelessly-charge-a-smartphone-safely-across-a-room/, 2018.

[3] Ellis, J. D., "Directed-Energy Weapons: Promise and Prospects," Technical report, Center for a New American Security, JSTOR, 2015.

[4] "Directed Energy Futures 2060: Visions for the Next 40 Years of U.S. Department of Defense Directed Energy Technologies," Technical report, U.S. Air Force Research Laboratory, 2021.

[5] Fox, R. L., *The Classical World: An Epic History of Greece and Rome*, Victoria, Australia: Penguin, 2006.

[6] Mills, A. A., and R. Clift, "Reflections of the 'Burning Mirrors of Archimedes.' With a Consideration of the Geometry and Intensity of Sunlight Reflected from Plane Mirrors," *European Journal of Physics*, Vol. 13, 1992.

[7] Wells, H. G., *War of the Worlds*, United Kingdom: William Heinemann, 1898.

[8] Fanning, W. J., "The Historical Death Ray and Science Fiction in the 1920s and 1930s," *Science Fiction Studies*, Vol. 37, No. 2, 2010, pp. 253–274.

[9] Longmire, C. L., "On the Electromagnetic Pulse Produced by Nuclear Explosions," *IEEE Transactions on Electromagnetic Compatibility*, Vol. EMC-20, No. 1, 1978, pp. 3–13.

[10] https://interestingengineering.com/atlas-i-the-cold-war-era-emp-test-facility, 2020.

[11] Baum, C. E., "Reminiscences of High-Power Electromagnetics," *IEEE Transactions on Electromagnetic Compatibility*, Vol. 49, No. 2, 2007, pp. 211–218.

[12] Baum, C. E., "A Time-Domain View of Choice of Transient Excitation Waveforms for Enhanced Response of Electronic Systems," in *International Conference on Electromagnetic Advanced Applications*, 2001, pp. 181–184.

[13] Li, Y., et al., "Simulation of High-Altitude Nuclear Electromagnetic Pulse using a Modified Model of Scattered Gamma," *IEEE Transactions on Nuclear Science*, Vol. 67, No. 12, 2020, pp. 2474–2480.

[14] Li, Y., et al., "Numerical Simulation of the Intermediate-Time High-Altitude Electromagnetic Pulse," *IEEE Transactions on Electromagnetic Compatibility*, Vol. 64, No. 5, 2022, pp. 1423–1430.

[15] Moran, S., "Historical Overview of Directed-Energy Work at Dahlgren," Technical report, Naval Surface Warfare Center, Dahlgren, VA, 2012.

[16] Radasky, W. A., C. E. Baum, and M. W. Wik, "Introduction to the Special Issue on High Power Electromagnetics (HPEM) and Intentional Electromagnetic

Interference (IEMI)," *IEEE Transactions on Electromagnetic Compatibility*, Vol. 46, 2004, pp. 314–321.

[17] Gardner, R. L., "Electromagnetic Terrorism. A Real Danger," in *11th Symposium on Electromagnetic Compatibility*, 1998.

[18] Giri, D. V., and F. M. Tesche, "Classification of Intentional Electromagnetic Environments," *IEEE Transactions on Electromagnetic Compatibility*, Vol. 46, 2004, pp. 322–328.

[19] Radasky, W. A., "The Threat of Intentional Interference (IEMI) to Wired and Wireless Systems," in *17th International Zurich Symposium on Electromagnetic Compatibility*, 2006.

[20] Clegg, B., *The Quantum Age: How the Physics of the Very Small has Transformed our Lives*, United Kingdom: Icon Books, 2014.

[21] https://www.ulsinc.com/learn/history-of-lasers, 2023.

[22] Einstein, A., "On the Quantum Theory of Radiation," *Physikalische Zeitschrift*, Vol. 18, 1917, pp. 121–128.

[23] Nauenberg, M., "Max Planck and the Birth of the Quantum Hypothesis," *American Journal of Physics*, Vol. 84, 2016, p. 709.

[24] Klein, M. J., "Max Planck and the Beginnings of the Quantum Theory," *Archive for History of Exact Sciences*, Vol. 1, No. 5, 1962, pp. 459–479.

[25] Bagayev, S., O. Krokhin, and A. Manenkov, "Pages from the History of Quantum Electronics Research in the Soviet Union," *Journal of Modern Optics*, Vol. 52, No. 12, 2005, pp. 1657–1669.

[26] Schawlow, A. L., and C. H. Townes, "Infrared and Optical Masers," *Physical Review*, Vol. 112, No. 6, 1958, pp. 1940–1949.

[27] Maiman, T. H., "Stimulated Optical Radiation in Ruby," *Nature*, Vol. 187, 1960, p. 494.

[28] https://laserfest.org/lasers/history/early.cfm, 2023.

[29] https://www.bbc.com/news/uk-politics-23518592, 2013.

[30] https://www.thedefensepost.com/2022/05/17/stryker-laser-mortars-drones/, 2022.

[31] https://www.rheinmetall-defence.com/en/rheinmetall_defence/index.php, 2023.

[32] https://edition.cnn.com/2014/12/11/tech/innovation/navy-laser-weapon, 2014.

[33] https://www.naval-technology.com/news/uss-portland-tests-laser-weapon-system/, 2021.

[34] https://www.naval-technology.com/analysis/navy-laser-weapon-systems/, 2019.

[35] https://www.reuters.com/article/idUSN1111660620100212?type=marketsNews, 2010.

[36] https://web.archive.org/web/20081211212241/http://www.af.mil/news/story.asp?id=123045745, 2007.

[37] https://www.lockheedmartin.com/en-us/news/features/2020/tactical-airborne-laser-pods-are-coming.html, 2020.

[38] https://www.epirusinc.com/products, 2023.
[39] https://www.army-technology.com/projects/leonidas-high-power-microwave-hpm-system-usa/, 2023.
[40] https://www.defensenews.com/land/2022/02/14/epirus-debuts-high-power-microwave-pod-for-drones/, 2022.
[41] https://militaryembedded.com/radar-ew/sensors/raytheon-emp-missile-tested-by-boeing-usaf-research-lab, 2012.
[42] https://phys.org/news/2012-10-boeing-champ-weapon-sci-fi-style.html.
[43] https://afresearchlab.com/news/afrls-drone-killer-thor-wins-popular-science-best-of-award/, 2021.
[44] https://www.raytheonmissilesanddefense.com/, 2023.
[45] https://www.latimes.com/archives/la-xpm-1989-07-18-mn-4091-story.html, 1989.
[46] https://www.popularmechanics.com/military/research/a28942057/neutral-particle-beam/, 2019.
[47] https://www.theguardian.com/technology/2006/oct/05/guardianweeklytechnologysection, 2006.
[48] https://www.bbc.com/news/newsbeat-10646540, 2010.
[49] https://genasys.com/lrad-2/maritime/, 2023.
[50] https://www.naval-technology.com/news/newslrad-to-deliver-acoustic-hailing-devices-us-navy/, 2013.
[51] https://jolt.richmond.edu/2019/11/30/lrad-the-sound-of-possible-excessive-force/, 2019.
[52] https://nsarchive.gwu.edu/briefing-book/intelligence-russia-programs/2022-09-15/moscow-signals-declassified-microwave, 2022.
[53] Elwood, J. M., "Microwaves in the Cold War: the Moscow Embassy Study and its Interpretation. Review of a Retrospective Cohort Study," *Environmental Health*, Vol. 11, 2012.
[54] Martinez, J. A., "The 'Moscow Signal' Epidemiological Study, 40 Years On," *Reviews on Environmental Health*, Vol. 34, No. 1, 2019, pp. 13–24.
[55] https://www.bbc.com/news/world-58396698, 2021.
[56] https://www.bbc.com/news/world-us-canada-55203844, 2020.
[57] https://www.cbc.ca/news/world/hong-kong-protest-lasers-facial-recognition-technology-1.5240651, 2019.
[58] Nielsen, P. E., *Effects of Directed Energy Weapons*, Washington, D.C.: National Defense University, 1994.
[59] Benford, J., and J. Swegle, *High-Power Microwave*, Norwood, MA: Artech House, 1992.
[60] Taylor, C. D., and D. V. Giri, *High-Power Microwave Systems and Effects*, Washington, D.C.: Taylor and Francis, 1994.
[61] Giri, D., *High-Power Electromagnetic Radiators: Nonlethal Weapons and Other Applications*, Cambridge, MA: Harvard University Press, 2004.
[62] Giri, D. V., R. Hoad, and F. Sabath, *High-Power Electromagnetic Effects on Electronic Systems*, Norwood, MA: Artech House, 2020.

[63] Anderberg, B., and M. L. Wolbarsht, "High Energy Laser (HEL) Weapons," in *Laser Weapons*, Boston, MA: Springer, 1992.
[64] Milonni, P. W., and J. H. Eberly, *Laser Physics*, Hoboken, NJ: Wiley, John Wiley & Sons, 2010.
[65] Zohuri, B., *High Energy Laser (HEL): Tomorrow's Weapon in Directed Energy Weapons Volume I*, Bloomington, IN: Trafford Publishing, 2014.
[66] Zohuri, B., *Directed Energy Weapons: Physics of High Energy Lasers (HEL)*, Cham, Switzerland: Springer, 2018.

CHAPTER

2

Contents

2.1 Probability and Statistics Basics

2.2 Principles of Stochastic Processes

2.3 Physics Preliminaries

References

Some Principles of Mathematics and Physics

The purpose of this chapter is to provide an overview of some of the foundational principles of mathematics and physics useful for understanding the results presented throughout the following chapters. It has been written from a minimalist perspective to provide the reader with a brief explanation of pertinent phenomena and to indicate where further information can be sourced. From a reference point of view, a useful guide on probability theory is [1] while a more practical guide to it is [2]. Simulation is an important tool for evaluating probabilities and a good reference on it is [3]. Stochastic process theory can be found in [4] while an excellent reference on the theory of queues is [5].

The approach in this book is to develop performance prediction based upon power density functions for HPRF DE systems and the irradiance function for HEL DE systems. From an RF perspective, useful references examining the power density and the radar-range equation include [6] and [7]. Another helpful reference, which not only contains a discussion of power densities but also electromagnetic waves and their propagation, is [8]. Additionally, one can consult [9] for principles of physics, while [10] and [11] provide useful discussions of electromagnetics. Specific references on laser physics are [12] and [13], while [14] can be consulted on the problems of laser beam propagation.

2.1 Probability and Statistics Basics

2.1.1 Fundamentals

Throughout the following, a random variable is a function on which a probability structure has been imposed to provide a model for an observed phenomenon. Occurrences of random phenomena are referred to as events and a probability measure is used to determine their likelihood. The set of all possible occurrences of a random phenomenon is referred to as a sample space, and is denoted Ω, while the set of all possible subsets of this sample space is referred to as a σ-field, and the elements of the latter are referred to as events.

A function $\mathbf{P}$ is a probability measure on Ω if it satisfies the requirements that:

1. $0 \leq \mathbf{P}(A) \leq 1$ for all events A;
2. $\mathbf{P}(\Phi) = 0$ and $\mathbf{P}(\Omega) = 1$, where Φ is the empty set;
3. If events A and B are disjoint then $\mathbf{P}(A \cup B) = \mathbf{P}(A) + \mathbf{P}(B)$.

A discrete probability distribution has a countable sample space such as the natural numbers or a subset of it. Suppose that $\Omega = \{s_1, s_2, \ldots, s_N\}$ is a set of N states then if p_j is the probability of state s_j ($p_j = \mathbf{P}(\{s_j\})$) then the collection $\{(s_j, p_j), j \in \{1, 2, \ldots, N\}\}$ is a general discrete distribution.

Random variables can be introduced in the above framework as follows. Suppose that X is a function acting on all subsets of Ω. Then using the probability measure $\mathbf{P}$ defined above, one can specify that $\mathbf{P}(X = s_j)$ is the probability that the random variable assumes state s_j, which is p_j by construction. The requirements imposed upon $\mathbf{P}$ enable one to determine probabilities of any collection of events.

Two events A and B are said to be independent if the probability of both of them occurring is multiplicative in the sense that $\mathbf{P}(A \cap B) = \mathbf{P}(A) \times \mathbf{P}(B)$.

Continuous probability distributions arise when it is supposed that a density for the random variable exists. The cumulative distribution function for a random variable X is defined to be

$$F_X(t) = \mathbf{P}(X \leq t), \tag{2.1}$$

where it is assumed that the random variable takes values on the real line. Then the random variable is called continuous if there exists a nonnegative function $f_X(t)$ such that

$$F_X(t) = \int_{-\infty}^{t} f_X(s)ds. \tag{2.2}$$

The function $f_X(s)$ is referred to as a probability density provided it is non-negative and integrates to unity.

Two random variables X and Y with a common support are said to have the same distribution if $\mathbf{P}(X \in A) = \mathbf{P}(Y \in A)$ for all sets A. This is often written in the form $X \stackrel{d}{=} Y$. In terms of distribution functions this means that $F_X(t) = F_Y(t)$ for all admissible t.

For a continuous distribution the mean measure of a random variable X is defined to be

$$\mathbf{E}(X) = \int_{-\infty}^{\infty} s f_X(s)ds, \tag{2.3}$$

with an analogous expression available for a discrete distribution. One can extend this definition to include other powers, referred to as moments of the random variable, so that for any $n \in \mathbf{N}$

$$\mathbf{E}(X^n) = \int_{-\infty}^{\infty} s^n f_X(s)ds. \tag{2.4}$$

The variance of a random variable is the average square deviation from its mean and is given by

$$\text{Var}(X) = \mathbf{E}\,(X - \mathbf{E}(X))^2 = \mathbf{E}\left(X^2\right) - (\mathbf{E}(X))^2. \tag{2.5}$$

A series of random variables $\{X_j, j \in \mathbf{N}\}$ is said to be independent and identically distributed (IID) if each has a common distribution and the independence property defined above applies to all events associated with the corresponding random variables. Hence for each i and j, $\mathbf{P}(X_i \in A) = \mathbf{P}(X_j \in A)$ and $\mathbf{P}(\{X_i \in A\} \cap \{X_j \in B\}) = \mathbf{P}(\{X_i \in A\}) \times \mathbf{P}(\{X_j \in B\})$, for any events A and B. Further details of random variables and their properties can be found in [1].

2.1.2 Distributions and Properties

Some specific examples of relevant probability distributions are now included, together with useful properties. A random variable X has an exponential distribution with parameter μ if its density is given by

$$f_X(t) = \mu e^{-\mu t}, \tag{2.6}$$

for $t \geq 0$ and zero otherwise. By applying (2.6) to (2.2) one can show that its cumulative distribution function is given by

$$F_X(t) = 1 - e^{-\mu t}, \tag{2.7}$$

also for the same domain as for the probability density function (2.6). Its mean is the reciprocal of μ, while its variance is the square of the reciprocal of μ. This distribution finds application to interarrival times in stochastic processes as well as a model for the area on a target that is vulnerable to radio frequency interference.

A random variable X has a gamma distribution with shape parameter α and scale parameter β if its density is

$$f_X(t) = \frac{\beta^\alpha}{\Gamma(\alpha)} t^{\alpha-1} e^{-\beta t}, \tag{2.8}$$

for $t \geq 0$.

The cumulative distribution function results from applying (2.8) to (2.2); in general a closed form expression for it does not exist, but must be evaluated through numerical integration. However, in certain cases it may be expressed as a series of functions. In particular, in the case where α is a positive integer the distribution reduces to an Erlang whose cumulative distribution function is given by

$$F_X(t) = e^{-\beta t} \sum_{i=\alpha}^{\infty} \frac{(\beta t)^i}{i!}. \tag{2.9}$$

Additionally, in this situation the gamma-distributed random variable can be decomposed into a sum of α IID exponential random variables with parameter β.

One of the simplest but most useful continuous distributions is that of a uniform or rectangular distribution. Suppose that R has a uniform distribution on the interval $[a, b]$. Then its density is given by

$$f_R(r) = \frac{1}{b-a}, \tag{2.10}$$

provided $a < r < b$ and is zero otherwise. Its cumulative distribution function is given by

$$F_R(r) = \begin{cases} 0 & r < a \\ \frac{r-a}{b-a} & a \le r \le b \\ 1 & r > b. \end{cases} \tag{2.11}$$

A uniform distribution generates random numbers in its support interval. It also provides a very useful way in which to generate realizations of any continuous random variable. In particular, if X is a continuous random variable with distribution function F_X then it can be shown that the random variable

$$F_X^{-1}(R) \stackrel{d}{=} X, \tag{2.12}$$

where R is uniformly distributed on the unit interval $[0, 1]$. The random variable on the left-hand side of (2.12) is well-defined because a distribution function is always increasing and so is invertible. As an example, suppose that X is exponentially distributed with distribution function (2.7). Then it can be shown that $F_X^{-1}(t) = -\frac{1}{\mu}\log(1-t)$. Hence (2.12) implies that $-\frac{1}{\mu}\log(1-R)$ has the same exponential distribution; since $1-R$ and R also have the same distribution one can simulate an exponential distribution through $-\frac{1}{\mu}\log(R)$.

Most of the distributions required throughout this book will be continuous, with one exception being the well-known Poisson model. A random variable Y whose support is the nonnegative integers has a Poisson distribution if its probability mass function is given by

$$p_Y(n) = \mathbf{P}(Y = n) = \frac{e^{-\mu}\mu^n}{n!}, \tag{2.13}$$

for all $n \in \mathbf{N}$. Its cumulative distribution function, for a subset A of the nonnegative integers is a sum of (2.13) over all elements of A, while its mean and variance are given by μ. Further details of these distributions can be found in [2].

2.1.3 Statistical Conditioning

Conditional probability is a useful tool in the evaluation of probabilities associated with performance metrics, and so it is necessary to provide an overview of such concepts. They will be required exclusively in the continuous distribution context, but the concepts are more readily understood from the discrete distribution setting, and in particular through the notion of sets. To introduce the concept of conditional probability, suppose that two events A and B are under consideration. Then the conditional probability that A occurs given B is defined to be

$$\mathbf{P}(A|B) = \frac{\mathbf{P}(A \cap B)}{\mathbf{P}(B)}, \tag{2.14}$$

provided $\mathbf{P}(B) > 0$. Hence conditional probability reduces the event sample space and the occurrence of an event is measured by how much it has in common with the given event, weighted by the likelihood of this event.

This idea has been extended to continuous distributions and provides a useful mechanism of evaluating probabilities. Suppose that X and Y are two jointly distributed continuous random variables with probability density $f_{(X,Y)}(x, y)$ such that the joint cumulative distribution function is given by

$$F_{(X,Y)}(x, y) = \int_{-\infty}^{x} \int_{-\infty}^{y} f_{(X,Y)}(u, v) du dv. \tag{2.15}$$

In the case where the two random variables are independent it follows that $f_{(X,Y)}(x, y) = f_X(x) f_Y(y)$ so that the joint density reduces to a product of the univariate densities.

The conditional probability density of X given $Y = y$ is defined to be

$$f_{X|Y=y}(x) = \frac{f_{(X,Y)}(x, y)}{f_Y(y)}, \tag{2.16}$$

provided $f_Y(y) > 0$. By rearranging (2.16) and integrating over the variable y one can show that

$$f_X(x) = \int_{-\infty}^{\infty} f_{X|Y=y}(x) f_Y(y) dy, \tag{2.17}$$

an expression which is of utility in the analysis to follow. In particular, by integrating (2.17) over x one can derive the expression

$$F_X(x) = \mathbf{P}(X \leq x) = \int_{-\infty}^{\infty} F_{X|Y=y}(x) f_Y(y) dy, \tag{2.18}$$

where $F_{X|Y=y}(x) = \mathbf{P}(X \leq x | Y = y)$ is a conditional cumulative distribution function.

Expressions for conditional expectations can also be derived, as illustrated in [1]. As a specific example, it can be shown that for two random variables X and Y that

$$\mathbf{E}(X) = \mathbf{E}(\mathbf{E}(X|Y)), \tag{2.19}$$

which can provide a useful mechanism to evaluate means when a dependent random variable is present.

2.2 Principles of Stochastic Processes

2.2.1 Fundamental Processes

A stochastic process $\{X_t, t \in \mathcal{T}\}$ is a series of random variables on a common support, indexed by the set $\mathcal{T}$. It is assumed that t refers to time and the values the process takes are states of the process. Hence a stochastic process, in very broad terms, is a model for the evolution of a system as it evolves over time. Specific stochastic processes arise as a probability structure is imposed on the system. The states of the process may be either discrete or continuous. Often in signal processing, a signal, which is represented as a stochastic process, is written simply as $f(t)$, and the two notations will be employed throughout the book.

One can associate with a stochastic process its mean measure, which is an auxiliary process given by $\{\mathbf{E}(X_t), t \in \mathcal{T}\}$. This mean function is often denoted $\mu_X(t) = \mathbf{E}(X_t)$. For the stochastic process defined above its autocorrelation function $R_X(s, t)$ is defined to be

$$R_X(s, t) = \mathbf{E}(X_s X_t), \tag{2.20}$$

while its autocovariance function is

$$C_X(s, t) = \mathbf{E}(X_s X_t) - \mu_X(s)\mu_X(t). \tag{2.21}$$

Observe that in the case where $s = t$ $R_X(t, t) = \mathbf{E}(X_t^2)$ and $C_X(t, t) = \text{Var}(X_t)$. The two processes (2.20) and (2.21) provide a means of quantifying the dependencies in the original stochastic process.

The stochastic process $\{X_t, t \in \mathcal{T}\}$ is called strictly stationary if the distribution of any subset of random variables defined through the process is time-invariant. Hence if $t_1, t_2, \ldots, t_N \in \mathcal{T}$ this definition implies that the joint distribution of the sequence $X_{t_1}, X_{t_2}, \ldots, X_{t_N}$ is exactly the same as that of $X_{t_1+\epsilon}, X_{t_2+\epsilon}, \ldots, X_{t_N+\epsilon}$ for any real ϵ such that each $t_j + \epsilon \in \mathcal{T}$. Such a stochastic process is usually referred to as being stationary for brevity.

A somewhat weaker version of stochastic process stationary is specified via the autocorrelation function (2.20). Specifically, a stochastic process is called wide-sense stationary if it satisfies the two conditions that

$$\mu_X(t) = \mu \text{ and } R_X(s,t) = R_X(s+\epsilon, t+\epsilon), \tag{2.22}$$

provided $s+\epsilon, t+\epsilon \in \mathcal{T}$. Hence condition (2.22) implies that the mean function of the process and its autocorrelation function are time-invariant. When discussing a wide-sense stationary stochastic process one often expresses the autocorrelation function in the form

$$R_X(s) := \mathbf{E}(X_t X_{s-t}) = \mathbf{E}(X_{t+s} X_t), \tag{2.23}$$

where the second equality is an immediate consequence of the second property in (2.22).

One of the primary stochastic processes of utility in the analysis to follow is a renewal process, which is a generalization of counting processes. The latter process can be specified, as in [4] by requiring that:

1. X_t is nonnegative for all t;
2. X_t takes values in the set of nonnegative integers;
3. $X_s < X_t$ whenever $s < t$;
4. Provided that $s < t$ the difference $X_t - X_s$ is the number of events that have occurred in the time interval $(s, t]$.

Such a process is said to have independent increments provided the number of events occurring in disjoint intervals are independent.

A Poisson process arises when the process is assumed to start at zero, with independent increments and provided the number of events occurring in a time interval has a Poisson distribution. In particular, for a Poisson process with mean arrival rate μ the latter requirement implies that

$$\mathbf{P}(X_{t+s} - X_s = n) = \frac{e^{-\mu t}(\mu t)^n}{n!}, \tag{2.24}$$

for all $s, t \geq 0$ and $n \in \mathbf{N}$.

A renewal process is a general process that enumerates events. Suppose that $\{X_n, n \in \mathbf{N}\}$ is a discrete indexed series of IID nonnegative random variables, and define the process $\{S_n, n \in \mathbf{N}\}$ through the conditions that

$$S_0 = 0 \text{ and } S_n = \sum_{j=1}^{n} X_j, \tag{2.25}$$

for $n \geq 1$. Viewing the series of X_n as interarrival times one can thus interpret the process S_n as the time of the nth event. Hence it is plausible to define a counting process

$$N(t) = \sup\{n \in \mathbf{N} : S_n \leq t\} \tag{2.26}$$

which is referred to as a renewal process.

The Poisson process defined previously arises with the choice that the interarrival times, given in the process $\{X_t\}$, are exponentially distributed with parameter μ.

In order to determine probabilities associated with the random variable $N(t)$ defined in (2.26) it is useful to observe the equivalence of events

$$\{N(t) \geq n\} \equiv \{S_n \leq t\}. \tag{2.27}$$

Therefore it follows that $\mathbf{P}(N(t) \geq n) = \mathbf{P}(S_n \leq t)$ and consequently

$$\mathbf{P}(N(t) = n) = \mathbf{P}(S_n \leq t) - \mathbf{P}(S_{n+1} \leq t). \tag{2.28}$$

Hence point probabilities of the counting process may be computed through consideration of the differences of appropriate complementary distribution functions of the underlying arrival process.

2.2.2 Queueing Theory

Queueing theory is concerned with analyzing the underlying structure in a process where customers arrive into a system, wait in line until the queue server is available, and then enter service. The customer leaves the system at the end of the service period. Such a system is modeled through an arrival process (such as a renewal process introduced previously), a process accounting for the delays customers experience while waiting for service and then the customer's service time when passed to the queue server. Throughout, it will be assumed that the customers enter the queue in a first come, first serve protocol and so cannot jump from one position to another in the queue.

An illustration of the queueing process is provided in Figure 2.1. It is supposed that arrivals into the queue are modeled by a renewal process as specified in Section 2.2.1. Hence interarrival times are given by the series $\{X_j, j \in \mathbb{N} := \{1, 2, \ldots\}\}$, which consists of nonnegative IID random variables. Then the time of arrival of the nth customer is given by (2.25). Suppose that the service time of the nth customer in the queue is modeled by a random variable Y_n, which is assumed to be independent of the arrival process. After customer n arrives into the queue it will experience a waiting time before it reaches the server. Suppose that the waiting time of the nth customer, also referred to as its delay, is given by D_n. This customer, after its arrival, spends a time $D_n + Y_n$ while in the queue and the next customer arrives at an interarrival time of X_{n+1} after customer n. Hence the delay that this new customer experiences can be shown to be given by

$$D_{n+1} = \begin{cases} D_n + Y_n - X_{n+1} & \text{if } D_n + Y_n \geq X_{n+1} \\ 0 & \text{otherwise,} \end{cases} \tag{2.29}$$

(see [4] for more details). Based upon (2.29), it follows that

$$D_{n+1} = \max\{0, D_n + Y_n - X_{n+1}\} \tag{2.30}$$

for $n \geq 1$ and $D_1 := 0$ since the first customer to arrive will not experience a delay but is immediately served. One can, in principle, solve (2.30) recursively, which can then be used to facilitate the determination of stationary distributions of the queue. However this will not be required for the performance prediction models developed in the following chapters.

What is of particular importance in the current context is to determine the number of arrivals that have been processed by the queue server. Towards

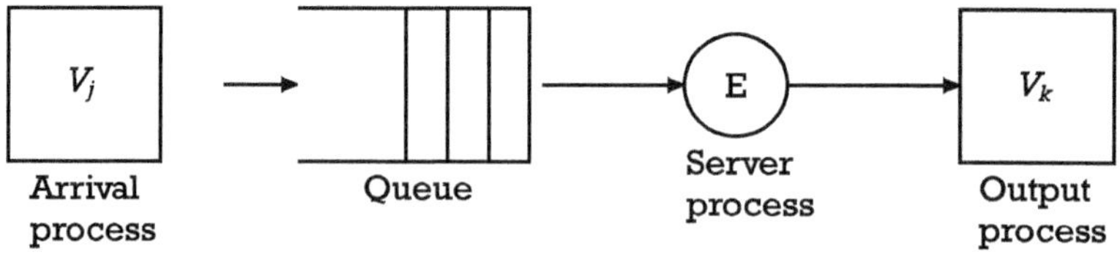

Figure 2.1 A single-server queueing theory model. Customers arrive into the system, at a rate of μ, and then are placed into the queue, in a first come, first serve protocol. Once a customer has been served it is considered a part of the output process. The latter process counts the number of customers that have been processed by the queue server by a given time.

this requirement, observe that the random variable $S_n + D_n + Y_n$ is the time at which the nth customer exits the queue, relative to time beginning at zero. Hence the random variable

$$M(t) = \sup_{n \in \mathbf{N}}\{S_n + D_n + Y_n \leq t\} \tag{2.31}$$

counts the number of customers that have passed through the queue by time t.

It is possible to construct the cumulative distribution function of (2.31) by observing the following result. If $M(t) \geq m$, for some $m \in \mathbf{N}$, then this will imply that $S_m + D_m + Y_m \leq t$. Similarly, if the latter holds then this will also imply that the former is also valid. Hence these two events are equivalent and therefore

$$\mathbf{P}(M(t) \geq m) = \mathbf{P}(S_m + D_m + Y_m \leq t). \tag{2.32}$$

Point probabilities of (2.31) can be produced by observing that

$$\mathbf{P}(M(t) \geq m) - \mathbf{P}(M(t) \geq m+1) = \mathbf{P}(M(t) = m). \tag{2.33}$$

The queueing model described above is usually described in Kendall's notation [15] as a G/G/1 queue; this notation indicates that a general arrival process is assumed, together with a general service time and that there is only one queue server. In the case where the arrival process is assumed Poisson, with exponentially distributed service times and one server, the queue is referred to as an M/M/1 model (G refers to general and M to a Markov process).

The above formulation of a queueing process assumes that customers remain in the queue until they are processed by the server. In reality, a customer may in fact decide to leave the queue while waiting for the server to become free. It is possible to reformulate the queueing model described above to reflect this situation, which will now be examined.

The situation where customers can exit the queue before service is referred to as reneging, and the following queueing theory model accounting for this has been adopted from [16]. Under this formulation it is again supposed that customers arrive into the queue through a renewal process as specified in the previous queueing model so that the interarrival times are given by $\{X_j, j \in \mathbf{N}\}$ with (2.25) again applicable for customer arrival times. It is assumed that customer n has a reneging time modeled through the random variable Z_n. This is the time that the nth customer will be prepared to wait in the queue before service. Alternatively, it can be interpreted as a customer's lifetime. If this customer's delay exceeds this random variable, the customer will exit the queue. Under the assumption that customer n reaches the server, this customer's service time is modeled through the random variable Y_n. If D_n is the delay that the nth customer perceives on arrival into the queue, and if one defines

$$U_n = \begin{cases} Y_n - X_{n+1} & \text{if } Z_n > D_n \\ -X_{n+1} & \text{if } Z_n \leq D_n \end{cases} \tag{2.34}$$

then the series of delays can be shown to be defined recursively through

$$D_{n+1} = \max\{0, U_n + D_n\} \tag{2.35}$$

for $n \geq 1$ with $D_1 = 0$ since the first arriving customer will not renege because it is served immediately.

Suppose that, as in the queueing model without reneging, we are interested in the number of arrivals that have been processed by the queue; in this case this includes both customers served as well as those that have reneged. Suppose that $M(t)$ counts the number of arrivals that have left the queue by time t. Then it can be shown that for any $n \in \mathbf{N}$

$$\mathbf{P}(M(t) \geq n) = \mathbf{P}(S_n + (D_n + Y_n)\mathbf{I}[Z_n \geq D_n] + Z_n\mathbf{I}[Z_n < D_n] \leq t), \tag{2.36}$$

where the indicator function $\mathbf{I}[x]$ takes the value unity in the case where condition x is true and is zero otherwise. One then may apply (2.36) to (2.33) to determine the point probabilities of $M(t)$.

It is also possible to explore networks of queues where customers can exit one queue prior to service and enter another, and where the customer can also renege. An excellent overview of networks of queues is [5]; a network of queues example will be featured in Chapter 6, and will be used to model the performance of two HEL DE systems.

2.3 Physics Preliminaries

A concise overview of relevant results from physics is now provided. As remarked previously, performance prediction will be based upon the power density function (for HPRF) and the irradiance function (for HEL) and so it is only necessary to introduce very basic concepts from applied physics. Hence Maxwell's equations, which provide the basis for the theory of electromagnetism and optics, is not overviewed. An understanding of the choice of wavelength and its propagation in space, as well as the transmission of waveforms to a target is of primary importance.

2.3.1 Wavelengths and Frequency

Electromagnetic waves propagate in free space and, if not subjected to attenuation, travel at the speed of light in a straight line, unless acted upon by a gravitational wave. These waves are characterized by their wavelength (λ, measured in meters) and frequency (f, measured in hertz (Hz)), which are related through the expression

$$\lambda = \frac{c}{f}, \tag{2.37}$$

where the speed of light in vacuo is $c = 299,792,458$ meters per second (m/s) (or approximately 300,000 km per second). Figure 2.2 provides an illustration of the electromagnetic spectrum. The two bands denoted *radio* and *microwave* are those which are referred to as radio frequency bands, and within it are the bands of relevance to HPRF DEWs. Infrared (IR) is the band of pertinence to HEL DEW systems. Hence the primary difference is that HPRF operates in meter to millimeter wavelengths, while HEL DEWs operate in the micron (10^{-6}) wavelength bands. Both DEWs apply electromagnetic energy to a target but are designed to achieve different effects. In the case of an HPRF effector, intense pulsed energy in the frequency range of 200 MHz to 5 GHz (wavelengths from 0.5996m reducing to 6 cm) have been shown experimentally to induce surface currents on the target that may then cause coupling into the system's electronics. By contrast a HEL DEW operates in significantly smaller wavelengths (and thus significantly larger frequencies)

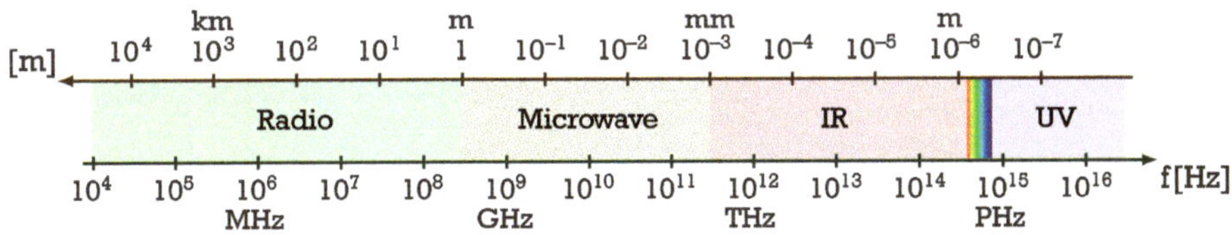

Figure 2.2 Illustration of the relationship between frequency and wavelength of electromagnetic waves, as well as the classification of the spectrum.[1]

1. This figure has been adapted by one produced by Izaak Neutelings from https://tikz.net/electromagnetic_spectrum/.

and applies focused energy to cause a thermal disruption. A HEL DEW with a wavelength of 10^{-6} m corresponds to a frequency of 0.2998 PHz.[2] As will be shown subsequently, one can characterize the effect of the DEW on a target through its power density function. This function provides a measure of the DEW's power concentration on the target relative to its distance from the DEW while accounting for attenuation effects on the waveform in the environment. Hence it is important to discuss these effects on the propagation of electromagnetic energy.

2.3.2 Propagation of Electromagnetic Energy

The electromagnetic field patterns that result from DE systems considered in this book can best be described as transverse electromagnetic (TEM) modes, where neither the electric nor magnetic fields are in the direction of propagation, as illustrated in Figure 2.3.

RF energy propagated in the Earth's atmosphere will be likely subjected to a number of distorting factors, in particular it is well known that at certain wavelengths RF transmissions will be attenuated by the presence of water vapor and gases. This attenuation can be significant in the presence of rain, fog, and clouds. The result of this, from an HPRF DEW viewpoint, is

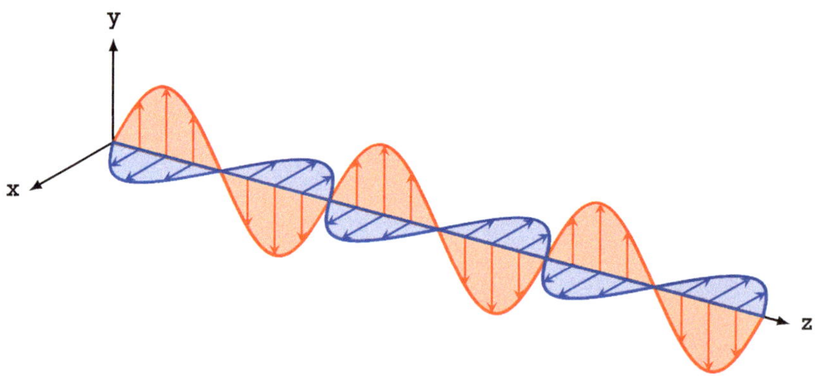

Figure 2.3 Propagation in transverse electromagnetic mode. Here the electric field is in the y-z plane, while the magnetic field is on the x-z plane. The direction of propagation is in the positive z-axis. (This figure has been adapted from https://tikz.net.)

2. PHz is petahertz, which is 10^{15} Hz.

a reduction in the effective power delivered to the target. Similarly, energy transmitted at wavelengths used for lasers are also subjected to degradation effects, including atmospheric attenuation and turbulence. In addition to the latter, a laser beam propagating in the atmosphere can be subjected to phenomena that can cause the beam to jitter when on the target. This results in an increase in the average spot size of the laser on the target and so reduces the amount of energy delivered to it. Hence it is important to account for such a distorting factor in the development of a performance prediction capability for DEWs. Such factors will be included in the development of power density functions introduced in subsequent chapters.

In order to illustrate some differences between HPRF and HEL DEWs in terms of propagation, it is necessary to discuss electromagnetic beam patterns. A useful reference on this from a radar perspective is [17]. When a resonator emits electromagnetic energy it will radiate it omnidirectionally unless it is designed to focus it into a specific direction. From an RF perspective a directive antenna will propagate its maximal energy in the desired direction but will inevitably dissipate energy in other directions. Hence plots of antenna radiation patterns will be similar to that illustrated in Figure 2.4. In this figure, an antenna's field strength is plotted as a function of the azimuth angle. It shows the mainlobe together with sidelobe radiation. The antenna's angular beamwidth is the width of the mainlobe measured from straight lines emanating from both sides of the mainlobe. The power of the antenna is produced by plotting the square of the field strength, and points where it hits zero are referred to as nulls. The radiation pattern in Figure 2.5 provides a plot of the corresponding antenna power associated with Figure 2.4, which also shows the points where there is 0- and 3-dB power loss relative to the maximum power level in the mainlobe.

When collimated electromagnetic radiation of wavelength λ is transmitted from an aperture with diameter d the resulting beam pattern has an angular divergence of approximately

$$\Delta = \frac{\lambda}{d}, \tag{2.38}$$

measured in radians. This result is under the assumption of perfect transmission of energy that is not attenuated by the environment; if such interference is considered then it is possible that this divergence is larger than the expression given in (2.38). The latter expression is a generalization of what is referred to as

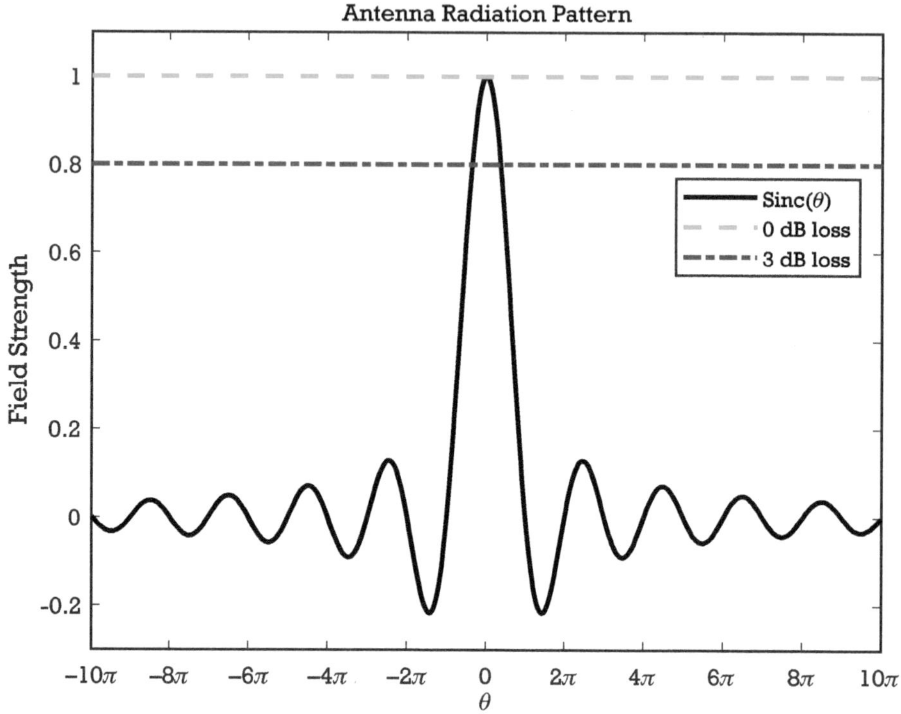

Figure 2.4 An illustration of an antenna radiation pattern, based upon the sinc function, showing field strength as a function of azimuth angle θ (in radians).

the 3-dB beamwidth in RF. To facilitate the interpretation of (2.38), consider the main beam of a resonator as illustrated in Figure 2.6. This figure plots the power in the mainlobe in the radiation pattern. The point N in this diagram is the position on the main beam where the power is maximum, while point M is the position where the power drops by 3 dB. The angle AOB is the angular width of the radiation pattern between the points on the radiation pattern that correspond to the 3-dB boundary. Hence the smaller this angular divergence the less the resultant energy diverges in free space.

Expression (2.38) can be used to clarify approximately the aperture differences between HPRF and HEL DEWs. To illustrate this, consider the case of a laser operating with a wavelength of $\lambda = 10^{-6}$ m, through an aperture of diameter $d = 10$ cm. Then the resultant beam has an angular divergence of approximately $\Delta = 10^{-5}$ radians. Now consider a microwave source

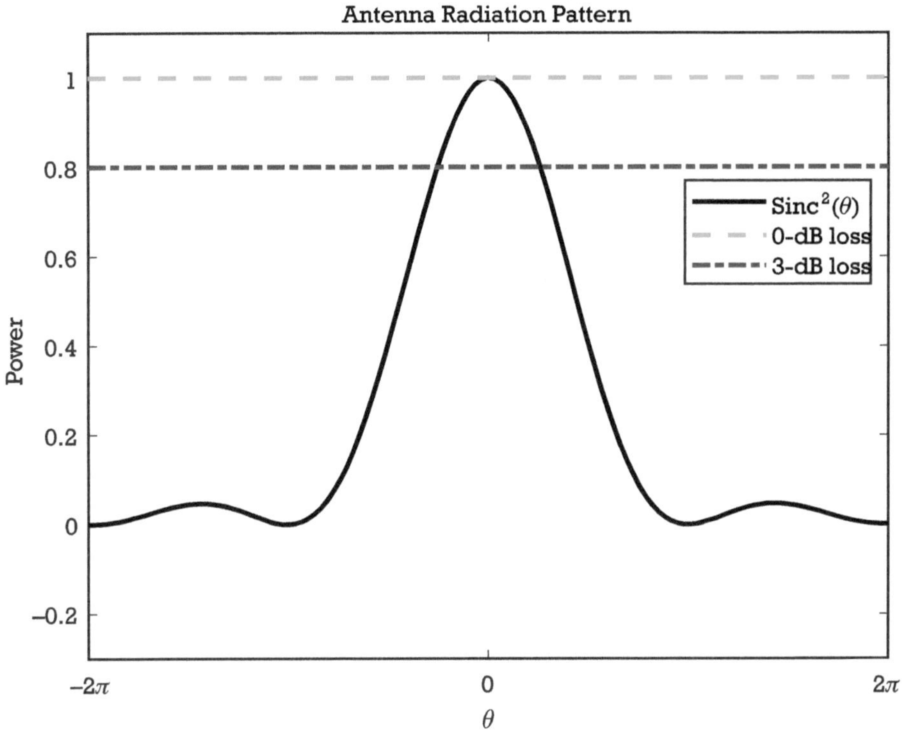

Figure 2.5 Antenna power associated with Figure 2.4 as a function of azimuth angle θ (in radians). The figure shows the 0- and 3-dB loss points of the main beam.

operating with a wavelength of $\lambda = 1$ cm. Then in order to achieve the same level of divergence this source requires an approximate aperture of diameter 1 km. Hence energy in the microwave spectrum diverges more rapidly with range than lasers, and for a microwave source to be beamed over a substantial distance necessitates either large apertures or shorter wavelengths. The latter requirement is not always feasible since, as pointed out in Section 2.3.1, there is a specific range of wavelengths where an HPRF DEW source can achieve its desired effect on a target.

2.3.3 Signals and Fourier Analysis

It is important to overview Fourier transforms and the representation of signals in the frequency domain, since this will be an integral component

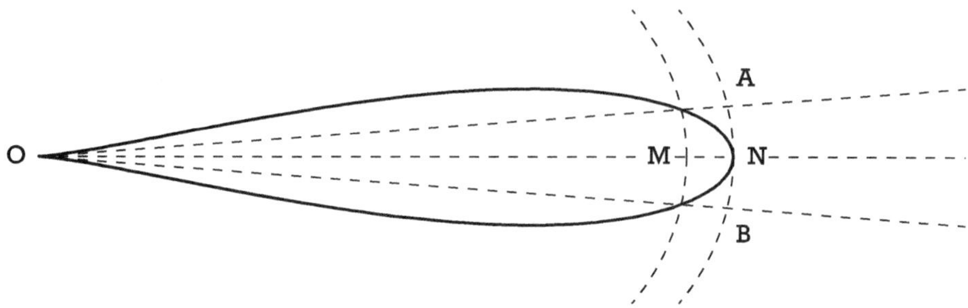

Figure 2.6 Determination of the 3-dB beamwidth angle. Point N is the position on the main beam where the power is maximum, while point M is the position where the power level drops by 3 dB. The angle AOB is the angular width of the radiation pattern between the points 3 dB down from the peak beam level.

in the analysis of wideband and ultrawideband signals to be introduced in the next chapter. Some of the discussion requires further properties of signals when viewed as stochastic processes. A suitable reference on Fourier transforms and their analysis is [18], while a reference that analyzes signals from a stochastic processes perspective is [19]. This section will follow the conventions used in signal processing, denoting a signal as a function of time. It is important to note that these signals can also be interpreted as a time-dependent stochastic process.

Suppose that $f(t)$ is a real-valued continuous time integrable signal. Then its Fourier transform is defined to be

$$\mathcal{F}(s) = \int_{-\infty}^{\infty} f(t)e^{-i2\pi st}dt, \tag{2.39}$$

where s is frequency measured in Hz. Observe that the Fourier transformation produces a function in the complex domain. The transformation (2.39) decomposes the time-dependent signal into a form from which its sinusoidal components can be analyzed. A signal such as a single sine wave has a Fourier transform that is a spike at its oscillation frequency and zero otherwise. A rectangular pulse waveform has a Fourier transform similar to the sinc function plotted in Figure 2.4.

A signal may also be recovered from the inverse Fourier transform so that

$$f(t) = \int_{-\infty}^{\infty} \mathcal{F}(s)e^{i2\pi st}ds. \tag{2.40}$$

The expected power of a real-valued signal will be defined to be

$$\mathcal{P} = \mathbf{E}(f(t)^2). \tag{2.41}$$

The power spectral density (PSD) of a signal describes its power distribution into frequency components and will be required in subsequent analysis. Suppose that $f(t)$ is a wide-sense stationary stochastic process with autocorrelation function $R(s)$. Then the PSD of this signal is the Fourier transform of its autocorrelation function. In particular, if S is the associated PSD then

$$S(x) = \int_{-\infty}^{\infty} R(s)e^{-i2\pi xs}ds. \tag{2.42}$$

The autocorrelation function can also be recovered from the PSD through an inverse Fourier transform, so that in particular

$$R(s) = \int_{-\infty}^{\infty} S(y)e^{i2\pi ys}dy. \tag{2.43}$$

It is informative to observe that by setting s to zero in (2.43) and applying (2.23) that

$$\int_{-\infty}^{\infty} S(x)dx = \mathbf{E}(f^2(t)), \tag{2.44}$$

Hence the expected power in a signal is the area under the PSD curve. In addition to this, if one constructs a bandpass filter from the original signal, so that it is the restriction of $f(t)$ for frequencies within a given band $[g_1, g_2]$, then one can show that the power in this signal is proportional to the area under the original PSD curve. Specifically if $g(t)$ is this restricted signal, and one is only focused on nonnegative frequencies, then

$$\mathbf{E}(g(t)^2) = \int_{g_1}^{g_2} S(x)dx. \tag{2.45}$$

Consequently, in order to determine the expected power in a signal over a frequency bin, one integrates the power spectral density over this frequency bin. Although this is possible in theory it is necessary to apply numerical integration in practice, which can be a challenge given that some PSD functions may not always be differentiable everywhere.

The applicability of Fourier analysis to understanding the structure of periodic signals with sinusoidal components is that it allows analysis to be undertaken in the frequency domain. This can be used to provide insights into otherwise difficult to analyze signals. Here a simple example is provided to illustrate the information that can be elicited from Fourier analysis. Towards this aim, consider a signal defined by

$$f(t) = A_1 \sin(2\pi f_1 t) + A_2 \cos(2\pi f_2 t) + A_3 \sin(2\pi f_3 t), \tag{2.46}$$

where $t \geq 0$, each A_j is an individual sinusoidal's amplitude, and each f_j is an oscillation frequency. Figure 2.7 provides a plot of this signal in the time domain. For this example $A_1 = 1$, $A_2 = 5$, and $A_3 = 10$, while $f_1 = 1.5$, $f_2 = 6.5$, and $f_3 = 10.5$ Hz. Under the assumption that one does not have a priori knowledge that the plot of this signal is from (2.46), one can conclude from inspection of the plot that the signal is composed of a number

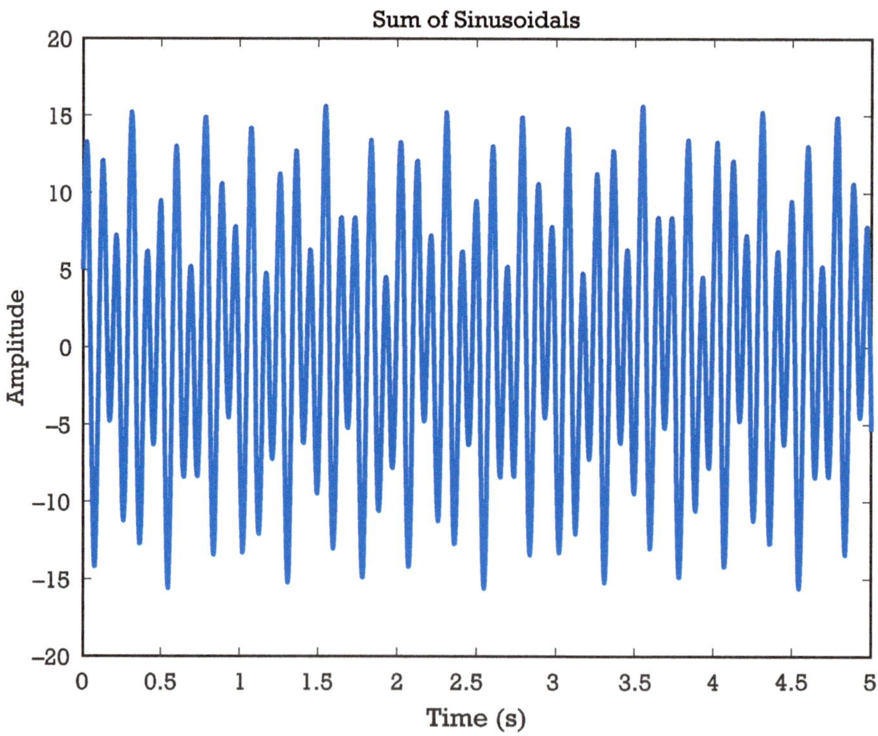

Figure 2.7 Plot of the signal defined by (2.46) as a function of time.

of sinusoidals. Based upon the data in Figure 2.7, the absolute value of the Fourier transform has been plotted in Figure 2.8, from which it can be concluded that the signal consists of three sinusoidal components, whose frequencies can be determined approximately from the figure.

Finally, Figure 2.9 plots the PSD for the data in Figure 2.7, showing the distribution of power in the signal per frequency. This plot also suggests that the underlying signal has only three frequency components, since it is almost zero except at three points. In view of (2.44), the area under this curve provides information of the distribution of signal power over a given frequency range. These techniques will be useful in the analysis of HPRF DE system performance prediction in the following chapter.

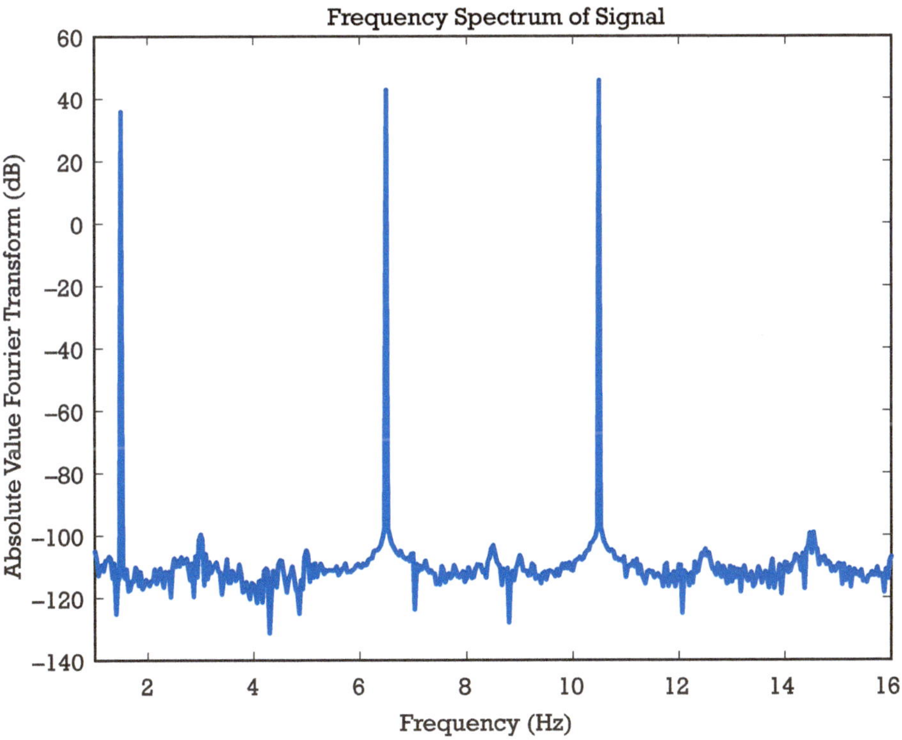

Figure 2.8 Absolute value of the Fourier transform based upon the data in the plot in Figure 2.7.

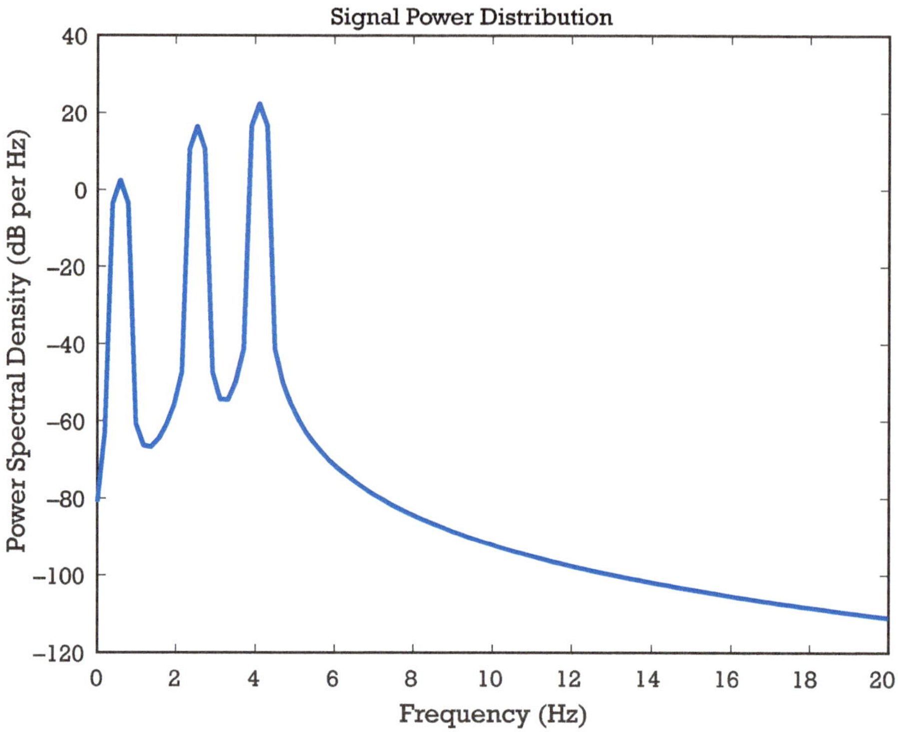

Figure 2.9 Example of a PSD for the data in Figure 2.7. The area under this curve for a given frequency bin provides an estimate of the signal power distributed over the corresponding frequency range.

References

[1] Billingsley, P., *Probability and Measure, 2nd Edition,* New York: John Wiley and Sons, 1986.

[2] Durrett, R., *Probability: Theory and Examples,* Belmont, CA: Wadsworth Publishing Company, 1996.

[3] Ross, S., *Simulation, 5th Edition,* San Diego, CA: Academic Press, 2013.

[4] Ross, S., *Stochastic Processes, 2nd Edition,* New York: John Wiley and Sons, 1996.

[5] Kelly, F. P., *Reversibility and Stochastic Networks,* Cambridge, U.K.: Cambridge University Press, 2011, https://www.statslab.cam.ac.uk/~frank/BOOKS/kelly_book.html.

[6] Levanon, N., *Radar Principles,* New York: John Wiley and Sons, 1988.

[7] Mahafza, B. R., *Introduction to Radar Analysis,* Boca Raton, Florida: CRC Press, 1998.

[8] Orfanidis, S. J., *Electromagnetic Waves and Antennas*, Piscataway, New Jersey: Rutgers University, 2016, https://www.ece.rutgers.edu/~orfanidi/ewa.

[9] Young, H., and R. Freedman, *University Physics with Modern Physics*, London, U.K.: Pearson Education Limited, 2019.

[10] Hammond, P., *Electromagnetism for Engineers: An Introductory Course*, Oxford, U.K.: Oxford University Press, 1997.

[11] Pollack, G. L., and D. R. Stump, *Electromagnetism*, Boston, MA: Addison Wesley, 2002.

[12] Milonni, P. W., and J. H. Eberly, *Laser Physics*, Hoboken, NJ: John Wiley & Sons, 2010.

[13] Shimoda, K., *Introduction to Laser Physics*, Springer Series in Optical Sciences, Boston, MA: Springer, 1984.

[14] Forbes, A., *Laser Beam Propagation: Generation and Propagation of Customized Light*, Boca Raton, FL: CRC Press, 2014.

[15] Kendall, D. G., "Stochastic Processes Occurring in the Theory of Queues and their Analysis by the Method of the Imbedded Markov Chain," *The Annals of Mathematical Statistics*, Vol. 24, No. 3, 1953, pp. 338–354.

[16] Stanford, R. E., "Reneging Phenomena in Single Channel Queues," *Mathematics of Operations Research*, Vol. 4, No. 2, 1979, pp. 162–178.

[17] Stimson, G. W., *Introduction to Airborne Radar*, Raleigh, NC: Scitech, 1998.

[18] Goldberg, R. R., *Fourier Transforms*, Cambridge, U.K.: Cambridge University Press, 2009.

[19] Pishro-Nik, H., *Introduction to Probability, Statistics and Random Processes*, Sunderland, MA: Kappa Research LLC, 2014, https://www.probabilitycourse.com.

CHAPTER

3

Fundamentals of HPRF Performance Modeling

Contents

3.1 An Overview of HPRF DEWs

3.2 Electronic Vulnerability Levels

3.3 HPRF Power Density Function

3.4 Narrowband and Wideband Effector Performance Modeling

3.5 Concluding Remarks

References

Our study of DEW systems begins with the case of HPRF and this chapter will provide an overview of HPRF principles and performance prediction fundamentals. The major objective is to derive the power density function for HPRF DEWs and to explore its characteristics. Initially it will be developed for the case of a continuous transmission of HPRF energy at a given wavelength. This will then be extended to the case where the HPRF DEW is operating in a wideband or ultrawideband mode, distributing the peak power over a given frequency range. There are a number of useful references on both HPRF DEW systems as well as their effects on targets, some of which include [1–4].

3.1 An Overview of HPRF DEWs

As remarked in Chapter 1, the existence of electromagnetic interference and its potential to damage an electronic circuit has been known for quite some time. Lightning strikes on aircraft, ships, or even terrestrial objects have also been observed to result in electrical currents flowing through vulnerabilities and consequently inducing damage into electrical systems [5, 6]. The resultant effects can be categorized as electromagnetic interference (EMI), and this can also occur when transmitting systems operate in similar bands, and hence it is of great concern to engineers to develop mitigation strategies against EMI [7]. This category of EMI is referred to as unintentional; as mentioned in the introduction, IEMI arose out of the realization that electromagnetic energy could be used for offensive and defensive purposes. Hence there is a wealth of literature documenting IEMI and its potential applications as well as the threat it poses to both military as well as civilian infrastructure [8].

An HPRF DEW applies significantly intense pulsed power to a target in order to achieve an effect through coupling electromagnetic waves into threat vulnerabilities [2]. Experimentally it has been demonstrated that intense electromagnetic waveforms, operating in the frequency range of 200 MHz to 5 GHz, are likely to induce effects in target systems [9]. The required intensity is peak power levels in excess of 100 MW and pulsed energy greater than 1 joule per pulse.

There are two ways in which these waveforms can be transmitted in order to achieve disruptive effects. A narrowband source generates coherent energy through an intense relativistic electron beam, and operates with a narrow bandwidth in the GHz frequency range [10, 11]. Such systems tend to require very large power sources for generation of the beam and its control [12]. By contrast, a wideband or ultrawideband source generates high peak power electromagnetic signals by direct excitation of an antenna system through high pulse voltages and a fast switching circuit, such as a Marx generator with a spark gap switch [13]. Such systems are often smaller in size than the narrowband counterpart, since they do not require the generation of an electron beam [14]. Typically they result in electromagnetic waves with a larger bandwidth than that of a narrowband source, and a reduction in output power, which is also distributed across the spectrum of frequencies. Further details of HPRF DE systems can be found in [15] or any of the textbooks listed in the previous section.

The objective of both categories of HPRF systems is to cause disruption to threats through coupling. As outlined previously, the two ways in which this can happen are through front and back door coupling. Front door coupling targets direct vulnerability points in a system where shielding is not possible, such as antennas and Wi-Fi ports, and requires a priori knowledge of the target's vulnerable operating frequency. This category of coupling can also be interpreted as providing a target system with false information [9]. A narrowband source is most likely to find application for such coupling, since it can be designed to target a well-known vulnerability frequency in specific classes of targets. Back door coupling occurs when energy is injected into a system through system vulnerabilities in physical cavities, wires, or computer ports, for example. Since it is generally not known which frequency will induce coupling, a wideband source has the potential to cause back door coupling, provided the field strength at the target is sufficiently intense. Such coupling often causes either transient upset or permanent damage [9].

One of the major benefits of an HPRF DEW is that target systems cannot be sufficiently shielded against them, since many of the vulnerabilities in systems require sensors or exposed antennas to perform their function [16]. However, this also means that the deployment of an HPRF DE system must be undertaken with due consideration of the possibility of unintentional EMI occurring in friendly systems. Examples include situations where an HPRF DEW operates near radar and tracking systems. Additionally, these considerations imply that the deployment of an HPRF DEW on specific platforms may be problematic. Hence dedicated land-based deployment of HPRF DEWs is likely to have the greatest advantage with minimal EMI issues.

There have been a number of studies documenting the application of HPRF DE to disruption of electronic systems in general. In particular, [17] is an early study of the impact of IEMI on wired and wireless systems, while [18] focuses on computer vulnerabilities. Specific studies examining effects at the component level include [19], while the impact of HPRF on modulation schemes has been documented in [20]. The susceptibility of commercial off-the-shelf (COTS) devices to HPRF has been examined in [21].

As remarked previously, it is of significant interest to apply HPRF DEWs to the problem of countering swarms of UAVs. Towards this objective there are quite a few studies that have examined the vulnerabilities of UAVs to HPRF attacks [22–25]. These vulnerabilities include their navigation, sensors, guidance, and control. Specific experimental studies undertaken to establish UAV vulnerabilities, in terms of HPRF DEW wavelength and field strength to achieve an effect, include [26] (navigation system attacks), [27, 28] (sensor system effects), and [29] (guidance and control disruption). In addition to this it is interesting to note that HPRF has also found application to the problem of forensic tracking of UAVs [30].

It is also worth remarking that HPRF DEWs have been demonstrated to achieve disruptive effects on other potential target systems. In particular it has been shown experimentally that an HPRF DE system can disrupt modern motors in cars [31, 32]. In such cases it is the electronic motor control, such as fuel injection, that can be affected and cause the system to shut down, and hence disrupt the vehicle.

The HPRF power density function, as well as the HEL irradiance function, is in units of power per surface area, and specifically measured in watts per square meter (W/m^2). However, in the HPRF and IEMI literature there is a preference to express field strength in units of volts per meter (V/m). Note that conversion from an electromagnetic field strength of x V/m to units in W/m^2 can be achieved through the simple relation given by

$$P(x) = \frac{x^2}{120\pi}. \tag{3.1}$$

Explicit examples of equivalent units can be found in Table 3.1.

3.2 Electronic Vulnerability Levels

The literature contains useful guides on not only components of electronic systems that are vulnerable to HPRF DEW attacks but also provides specifications on the required field strengths at operating frequencies that can achieve specific coupling effects. In this section field strength will be in V/m to be consistent with the literature cited. The reader can refer to Table 3.1 for conversion to W/m^2 units.

Table 3.1 Volts per Meter and Watts per Square Meter Equivalent Pairs, Based Upon the Transformation (3.1)

Volts per Meter (V/m)	*Watts per Square Meter (W/m^2)*
50	6.6315
55	8.0241
100	26.5258
300	238.7324
500	663.1456
1k	2.6526×10^3
15k	5.9683×10^5
24k	1.5279×10^6

A useful guide on vulnerability thresholds for electronic systems is [17], which not only examines computer system vulnerability levels but also investigated disruption of electronic control in modern motor vehicles. These threshold levels can be used to extrapolate those for airborne systems, and they indicate the required electromagnetic field strength to cause various disruptions. To illustrate, it was reported that motor control was disrupted with an electromagnetic field strength of 500 V/m, while permanent damage to the engine's control could be achieved with an electromagnetic field strength of 15 kV/m at 1.3 GHz, and at 24 kV/m at 2.86 GHz.

In terms of electromagnetic field strengths to disrupt sensor systems, [28] provides relevant experimental results. This study examined three commercially available sensors: a magnetometer, a barometric pressure sensor, and a current sensor. Examples of reported results include status and serial errors in the magnetometer with a field strength of 50 V/m for waveforms in the frequency range of 300 to 500 MHz and also for signals in the frequency range of 1.8 to 1.9 GHz with an electromagnetic field strength of 300 V/m. The barometer experienced various errors over a wider variation of field strengths and frequencies. To illustrate, for a field strength of 100 V/m, disruptions were observed at frequencies ranging from 1.2 to 7.4 GHz.

In terms of UAV vulnerabilities it is useful to note that [22] demonstrates that waveforms in the 100 MHz to 2 GHz frequency band have the greatest impact on UAVs. Examples of observed disruptions include variations in the UAV's vertical speed with a field strength of 55 V/m, deviations in its yaw angle with 100 V/m field strength, and loss of gimbal data at the ground station with a similar field strength.

These results provide rules of thumb in terms of the specification of minimal HPRF DEW settings for counter swarm UAV applications.

3.3 HPRF Power Density Function

The HPRF power density function can be viewed as a one-directional version of the well-known radar range equation, where the latter expresses the maximum radar detection range in terms of antenna gain and aperture effective area [33]. It is one-directional in the sense that while a radar is concerned with the returned waveform, which is then used for characterization purposes of a threat, an HPRF DE source is not designed to receive and process reflected waveforms. Hence the derivation of the power density, which provides a measure of the HPRF power incident on a target, follows that of the radar range equation.

It is necessary to introduce the concept of an antenna's gain to facilitate the reader's understanding of the development. Antenna gain is defined to be the ratio of power from the transmitter per unit of solid angle emitted in a given direction to the power of an isotopically radiating antenna over the same solid angle. Antennas will consequently experience gain in almost all directions, as can be observed in the example in Figure 2.4. The antennas of relevance in the current context will attempt to focus energy in one particular direction, through a highly directive antenna, and the gain in the center of the mainlobe in an antenna radiation pattern is taken as a metric for antenna performance. The mathematical expression for gain is given by

$$G_A = \frac{4\pi A_R}{\lambda^2}, \tag{3.2}$$

where A_R is the effective area of the aperture and λ is the operational wavelength of the HPRF DE source.

As explained previously, the power density is a measure of the antenna power distributed over a surface area, and so is measured in watts per unit area. From the point of view of an omnidirectional antenna radiating in free space, the corresponding radiation pattern is spherical and hence the peak power density can be measured as the ratio of peak transmitted power to the surface area of a sphere. Hence the power density is given by

$$I_1(R) = \frac{P_T}{4\pi R^2} \tag{3.3}$$

where P_T is the transmitted peak power (in watts) and R is the distance between the radiating source and the target. Expression (3.3) needs to be adjusted to account for power density focused by an antenna on a target in a given direction. Hence if the corresponding directive antenna has maximal gain of G_A then the power density focused on the target is therefore

$$I_2(R) = \frac{P_T G_A}{4\pi R^2}. \tag{3.4}$$

One can now apply the gain (3.2) to (3.4) to produce the peak power density in terms of antenna effective area, and so

$$I_3(R) = \frac{P_T A_T}{\lambda^2 R^2}. \tag{3.5}$$

To produce a final expression, suppose that at time t the distance between the aperture and target is $R(t)$, measured in meters (m). Then the peak power density can be written

$$I_4(t) = \frac{P_T A_T}{\lambda^2 R(t)^2}, \tag{3.6}$$

which will be referred to as the power density throughout for brevity.

As an aside, a derivation of a power density function for HPRF effects modeling has been introduced in [34], along similar lines of reasoning as introduced in this chapter. However, the result in [34] has been designed so that it is integrated into threat detection, for application to defense of a series of AFVs facing a single threat. It is important to note that it is not as general as the results presented in this chapter.

It is possible to rearrange the power density (3.6) to provide the maximum range from the antenna to a point where the power density reaches a given level. In particular it can be shown that

$$R(t) = \sqrt{\frac{P_T A_T}{\lambda^2 I_4(t)}}. \tag{3.7}$$

Some illustrations of the results encapsulated in (3.6) and (3.7) are now provided. In the first case, Figure 3.1 provides examples of (3.6). For the case where the HPRF source power is $P_T = 10^9$ W (1 GW) and with an effective antenna area of 2 square meters (m^2), the figure plots (3.6) for the four frequencies of 200 MHz, 500 MHz, 1 GHz, and 5 GHz for distances of up to 1 km. Since each plot corresponds to a different wavelength, but common antenna effective area, each has a different gain. These are 10.4866 dB (200 MHz), 18.4454 dB (500 MHz), 24.4660 dB (1 GHz), and 38.4454 dB (5 GHz), respectively. The figure demonstrates that at higher frequencies (or shorter wavelengths) the power density on a target will be sufficiently large. In fact it can be observed that for the case of 5 GHz the power density on a target is of the order of 10^5 W/m^2 at a range of 1 km, which is a sufficiently strong field strength to disable the electronics in many systems.

Figure 3.2 provides an example of (3.7) for the case of an HPRF source operating at a frequency of 1 GHz, with a peak power of $P_T = 10^{12}$ W and with an antenna effective area of 1 m^2. To illustrate the way in which this result may be used to produce rules of thumb, suppose that one was interested in disabling a target that required a field strength of 10^6 W/m^2 to guarantee a

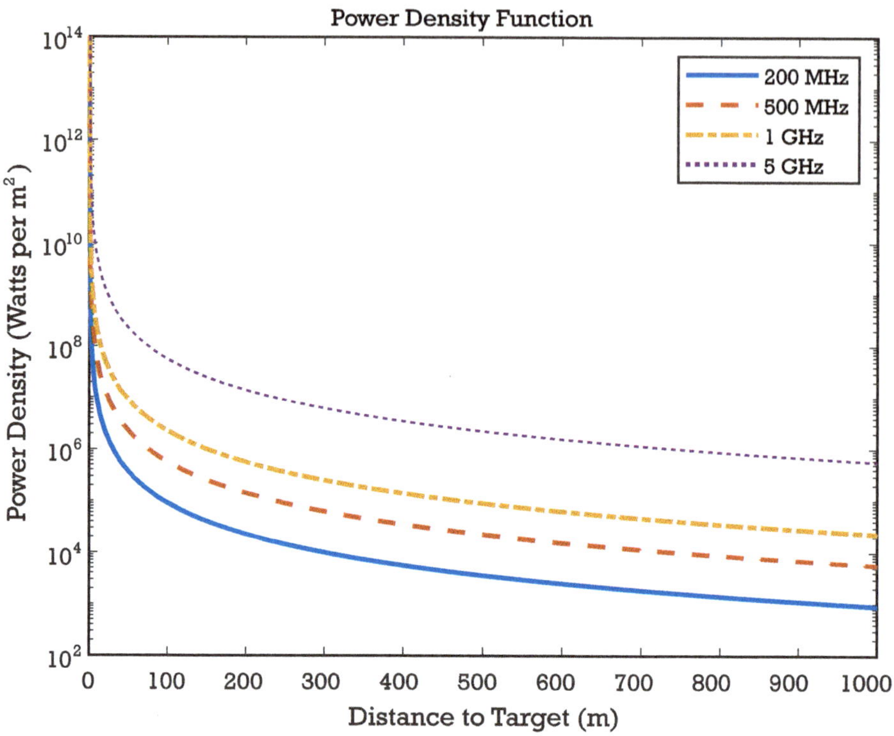

Figure 3.1 Four examples of the power density function (3.6) with a peak transmitter power of 1 GW.

disruption. Then from this figure one can conclude that this will happen at a maximum range of approximately 3 km. Similarly, the same figure enables one to conclude that at roughly 5 km the power density on a target would be 10^4 W/m^2. These results need to be interpreted carefully though, since the performance is under idealized conditions, since the effects of atmospheric attenuation are not included in the power density function (3.6).

Although the power density (3.6) enables one to undertake some preliminary analysis of HPRF DE performance, it lacks an important feature. When RF energy is pulsed toward a target it is not certain that all the energy received at the target will in fact be effective in achieving disruption. Hence one can introduce a factor into (3.6) to account for what is often termed the coupling cross section of the target. This is analogous to the application of

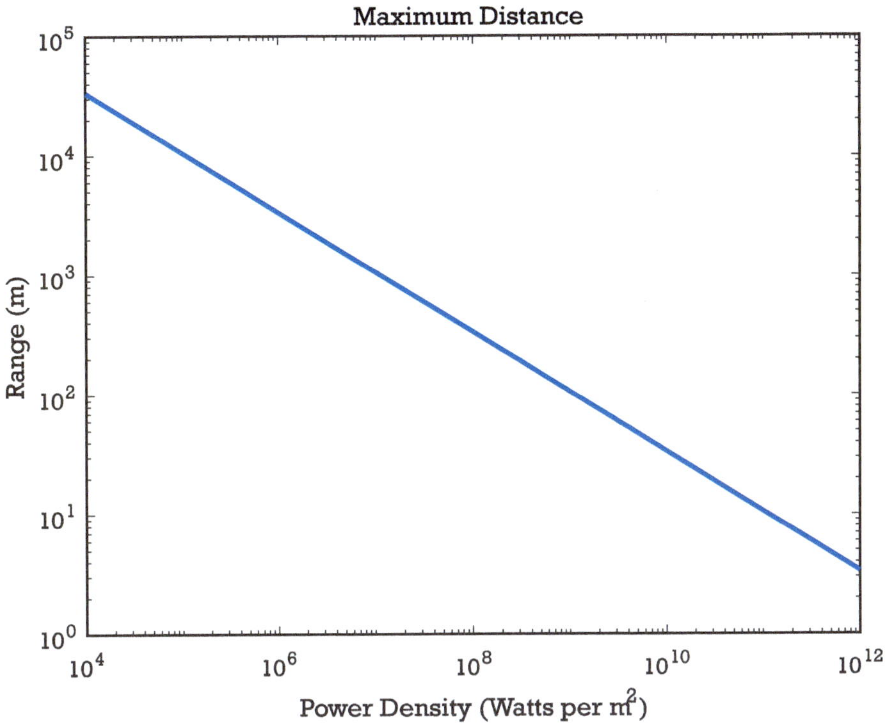

Figure 3.2 Maximum ranges for given power density levels, for the case of an HPRF effector operating at 1 GHz with peak power of 10^{12} W.

radar cross section (RCS) to characterize a target from an electromagnetic point of view. Hence suppose that A_R is the coupling cross section of the target. Then the final form of the power density function is given by

$$I(t) = \frac{P_T A_T A_R}{\lambda^2 R(t)^2}. \tag{3.8}$$

The power density in (3.8) is exactly that given by the well-known Friis' transmission formula [35], except in the interpretation adopted here A_R will be considered stochastic in nature. Hence specific statistical models may be applied to A_R.

Towards this objective, suppose that A_R has distribution function $F_{A_R}(\cdot)$, and suppose that the parameters P_T, A_T, λ, and the distance $R(t)$ are deterministic. Then based upon (3.8) it can be shown that the power density function has distribution function

$$\mathbf{P}(I(t) \leq x) = F_{A_R}\left(\frac{x\lambda^2 R^2(t)}{P_T A_T}\right), \tag{3.9}$$

for any real $x \geq 0$. Hence if one specifies a distribution for the coupling cross section then one can further simplify (3.9).

As an example, suppose that A_R is modeled through an exponential distribution with parameter μ. This choice has been motivated by what is known as a Swerling I target model used in the theory of radar detection [36]. Hence under this assumption (3.9) reduces to

$$\mathbf{P}(I(t) \leq x) = 1 - e^{-\frac{\mu x \lambda^2 R^2(t)}{P_T A_T}}. \tag{3.10}$$

The form of the distribution function in (3.10) implies that the power density function has an exponential distribution. The complement of (3.10) can be used to determine the likelihood that the power density function achieves a given minimum power level on a target at a given distance. To illustrate this, consider the case of an HPRF source operating at 1 GHz with a power level of 1 GW through an antenna with effective area of 1 m^2. Suppose that the mean of the A_R is unity. Then Figure 3.3 contains four plots of the complement of (3.10), for the cases where the minimum power density level x is 10^4, 10^5, 10^6, and 10^7 W/m^2, respectively. The figure demonstrates the fact that as this minimum level increases the likelihood of HPRF DE disruption of the target occurs at shorter distances from the effector, and to achieve disruption at distances further from the effector the minimum level of disruption must be smaller.

It will also be important to consider the power density function when it is applied over an observational window of time, which is a period when the target is illuminated by an HPRF effector. Over an observational window $[t_0, t]$ $(t_0 < t)$ the power density is defined to be

$$I_{\{t_0\}}(t) = \int_{t_0}^{t} \frac{P_T A_T A_R}{\lambda^2 R(s)^2} ds. \tag{3.11}$$

Suppose that A_R has an exponential distribution with parameter μ then one can show that

$$\mathbf{P}(I_{\{t_0\}}(t) \geq x) = e^{-\frac{\mu x t_0 t \lambda^2}{P_T A_T [t - t_0]}}, \tag{3.12}$$

for $x \geq 0$. To illustrate the evaluation of (3.12) suppose that an HPRF DEW operates at a frequency of 1 GHz with 1 GW of power through a

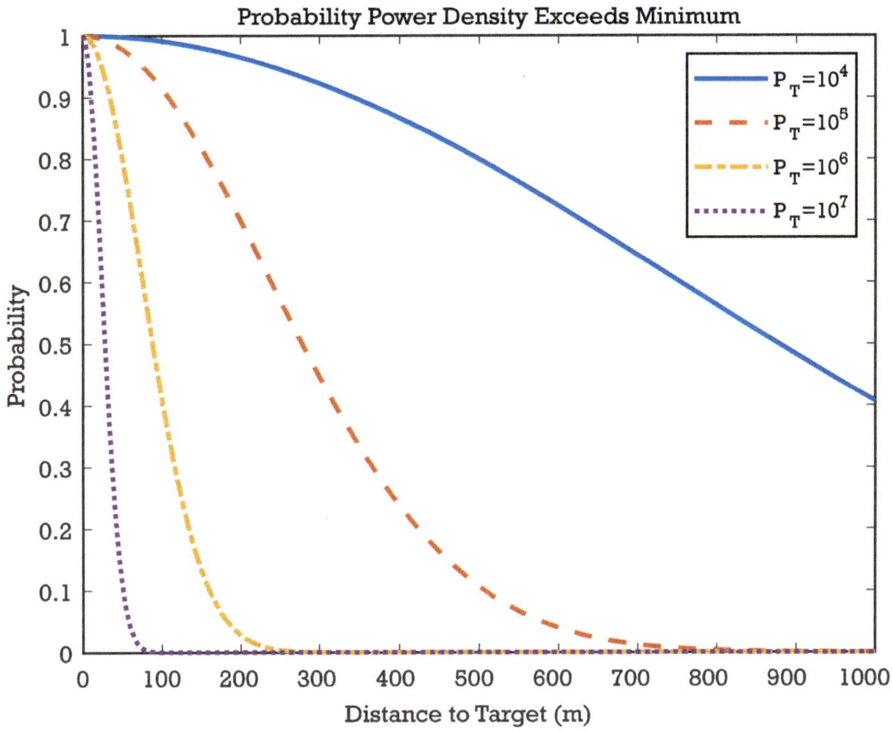

Figure 3.3 Plots of the probability that the power density exceeds minimum levels, as indicated through the legend values.

directive antenna with $A_T = 2$, and that $\mu = 1$. Then the corresponding gain is $G = 24.466$ dB and Figure 3.4 provides a plot of the complementary distribution function of the power density (3.11) over four observational windows: [1, 1.1], [1, 1.5], [1, 2], and [1, 5] respectively. The figure demonstrates that as the observational window increases the complementary distribution function will have larger probabilities as expected, since the HPRF DEW is spending more time on creating the desired energy increase on the target.

3.4 Narrowband and Wideband Effector Performance Modeling

The power density functions developed in the preceding section focused on the case where there is a unique HPRF DEW source power operating at a

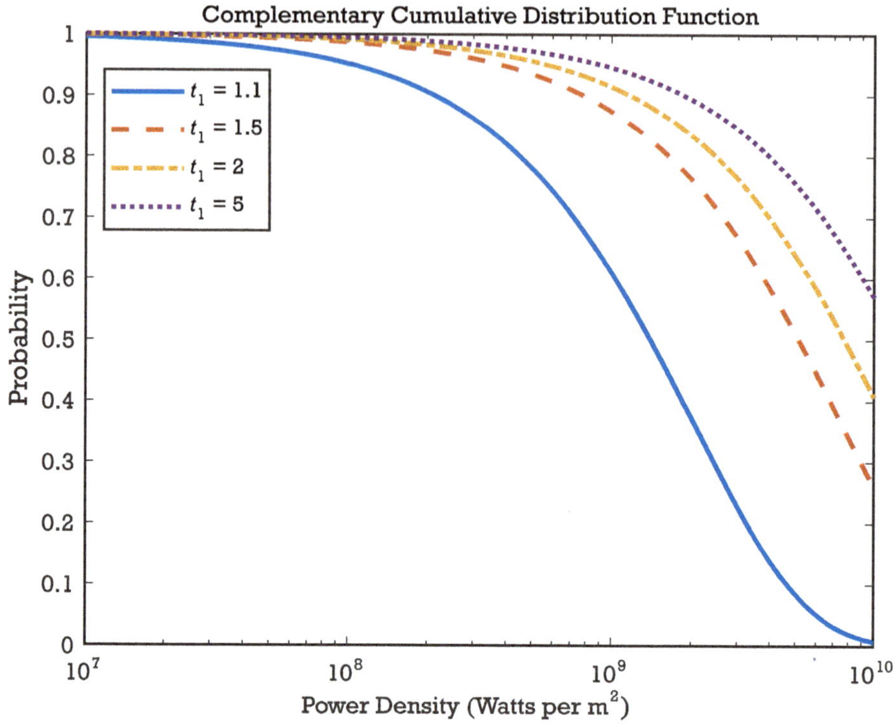

Figure 3.4 The complementary distribution function (3.12) over a series of four observational windows.

given wavelength. The purpose of this section is to investigate how this can be extended to provide a power density function for the modeling of the situation where the HPRF DEW source's power is distributed over a range of frequencies. As will become apparent, this is a nontrivial exercise, and it will be necessary to propose some approximate solutions. In the first place, it will be necessary to clarify the differences between narrowband, wideband, and ultrawideband HPRF sources.

A narrowband source operates with at most a 10% bandwidth around its center frequency, where the latter is usually in the GHz frequency range. Wideband and ultrawideband systems operate in the frequency range of 0.01 to 2 GHz, where a wideband system operates with a bandwidth of between 10% and 25% around its center frequency, while an ultrawideband system extends this bandwidth to at least 25%. Figure 3.5 provides an illustration

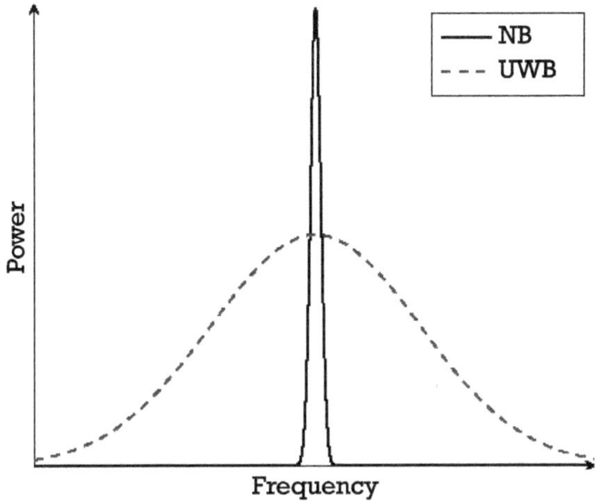

Figure 3.5 Narrowband (NB) and ultrawideband (UWB) signals showing the power distributed over frequency range.

of the differences between a narrowband and ultrawideband signal in terms of the way in which the power is distributed over the permissible frequency ranges. From an HPRF DEW effects perspective, since the target's vulnerable frequency range is unknown, an ultrawideband effector is likely to match at least one such frequency, but the effector's power at this frequency will be less than that of a narrowband source.

The mathematical models developed thus far for the power density function on a target can be viewed as modeling an essentially narrowband source, although there has not been strict adherence to the definitions provided above. The following analysis will be concerned with proposing a way in which an ultrawideband signal can be analyzed in terms of an effector's power density function.

Suppose that at time t a signal is transmitted to an intended target, with a power of $P(t)$ watts, and at a wavelength of $\lambda(t)$ meters. Assume that at time t the target is at a distance of $R(t)$ meters away, and that the HPRF DEW's effective area remains uniform (A_T) but suppose that the target's coupling cross section varies with wavelength, so that it is $A_R(\lambda(t))$. Then the power density on the target at time t is

$$I(t) = \frac{P(t)A_T A_R(\lambda(t))}{\lambda^2(t)R^2(t)}. \tag{3.13}$$

The 2-tuple $(\lambda(t), P(t))$ can be selected to model the characteristics of an ultrawideband source, while $A_R(\lambda(t))$ can be chosen to reflect the expected likelihood of coupling into a target at different wavelengths.

Although (3.13) provides a generic way in which to analyze the power density on a target for a wideband or ultrawideband source it does not have usefulness in practice. A major problem with applying (3.13) is that, as will be evident in the following chapter, one must compare (3.13) to a wavelength-dependent threshold. The latter measures the point at which the power delivered to the target is sufficient to cause coupling. However, this will vary with wavelength and so it would be difficult to apply (3.13) without specifying a cascade of thresholds to which it must be compared to on a wavelength to wavelength basis.

Before proposing an alternative to (3.13) it will be insightful to consider a classical ultrawideband signal and to investigate the distribution of power across its frequency spectrum.

3.4.1 Example: A Damped Sinusoidal Signal

The quintessential example of an ultrawideband signal is a damped sinusoidal function, of which the following is a typical case:

$$f(t) = Ae^{-\zeta t}\sin(2\pi f_0 t), \tag{3.14}$$

where $t \geq 0$, A is the signal amplitude, f_0 is the oscillation frequency, and ζ is the exponential decay parameter. Such a signal is not permissible in wideband communications but has great utility in the context of HPRF DEW [37, 38]. A specific parameterized example of this signal is provided in Figure 3.6. As a pulsed signal for HPRF DEW coupling it is necessary to have a short pulse length with a fast rise time. Hence the pulse width is 10^{-6} seconds (s), with an amplitude of $A = 66.5595$ dB, $\zeta = 5 \times 10^6$, and $f_0 = 5$ MHz. The signal has been sampled at 256 points. Recall that one can measure the signal power through the integral of the square of the signal over its pulse duration. Based upon this, the signal defined in (3.14) can be shown to have a power level of 10^6 W.

A pulse train based upon this signal can be produced with a given pulse repetition rate. From a Fourier analysis perspective, one can examine a single pulse to gain an understanding of the train's frequency distribution.

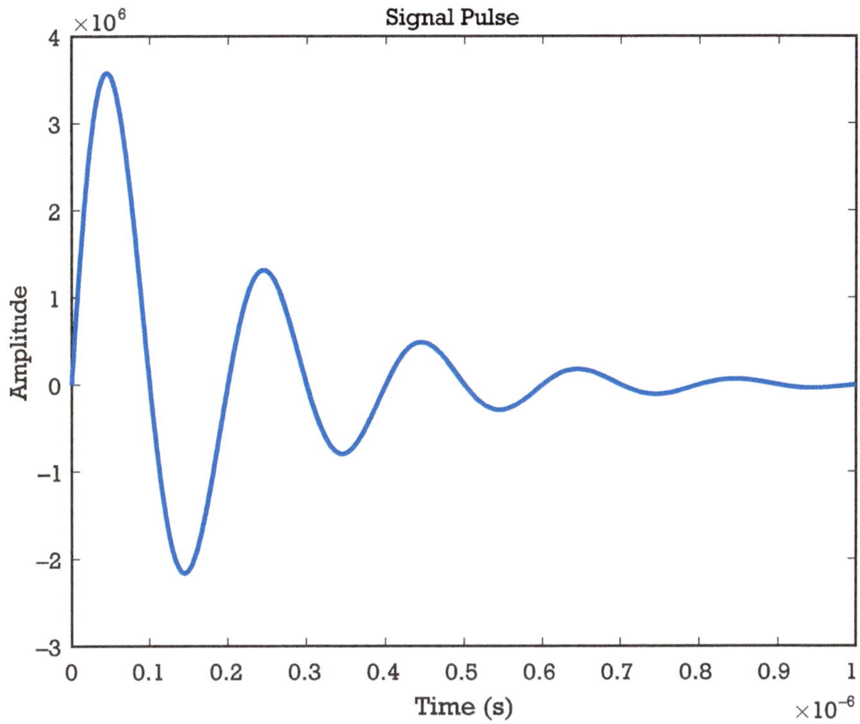

Figure 3.6 A damped sinusoidal pulse signal, which has a short pulse width of 10^{-6} s and an amplitude chosen so that the overall signal power is 60 dB watts.

Figure 3.7 provides a plot of the absolute value of the Fourier transform of this signal, which is given by

$$\mathcal{F}(s) = \left| \frac{2\pi f_0 A}{\zeta^2 - 4\pi^2(s^2 - f_0^2) + 4\pi i \zeta s} \right|. \tag{3.15}$$

The plot in Figure 3.7 demonstrates the distribution over frequencies of sinusoidal components inherent in the signal (3.14). As can be observed, with an approximate center frequency of 5 GHz the frequencies range from 0 to 10 GHz.

It is of interest to determine how the signal's power is distributed across its frequency range. As discussed previously, this can be determined

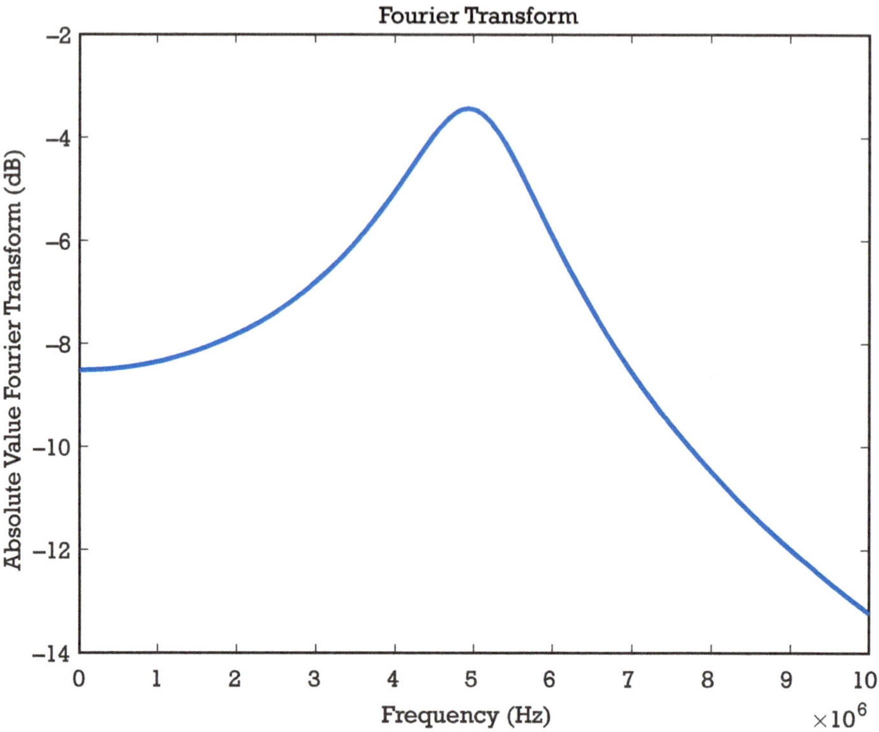

Figure 3.7 The absolute value of the Fourier transform of the signal (3.14), demonstrating the signals inherent frequency components.

by estimating areas under the power spectral density, which is shown in Figure 3.8. From the point of view of (3.13) it will be somewhat difficult to estimate exact power levels for a given frequency since, as Figure 3.8 indicates, there are regions where a small frequency window may result in difficulty in approximating the area under the curve via numerical methods such as Simpson's rule or Trapezoidal approximation. This is also apparent from the differences in scales in the figure, noting that the area will need to be estimated from this curve but in standard units on the vertical axis. Despite these concerns, numerical integration produced the results shown in Table 3.2, where over given frequency windows or bins the relative power has been estimated. These choices of frequency bins have been based upon success in determining a numerical result, and also in an attempt to identify

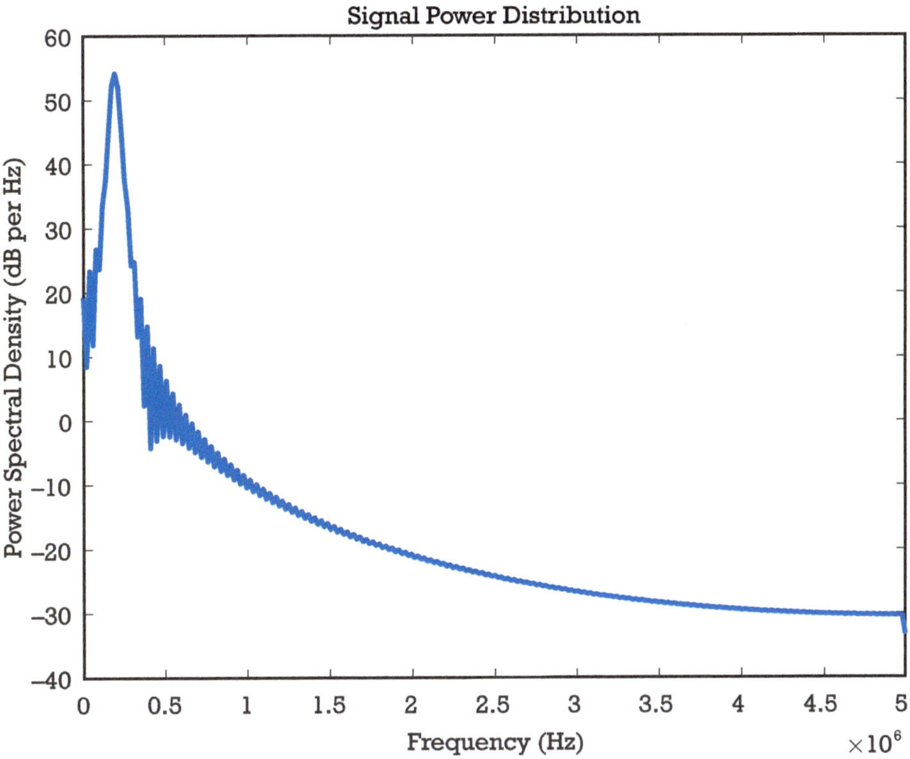

Figure 3.8 The power spectral density of the signal under consideration, plotting the power per frequency as a function of frequency.

the intervals where most of the power in the signal is concentrated. As can be observed from the data in Table 3.2 most of the signal power is in the interval $[10^2, 10^6]$ Hz. In view of the results in Table 3.2 one can then determine the power in any of the given frequency bins. Given the total power is 10^6 W, in the frequency bin $[10^5, 10^6]$ the proportion of power is 6.298×10^5 W, for example.

From an effects prediction perspective, note that since the power in the window $[0, 10^4]$ is insignificant, one can focus on performance prediction based upon frequencies in the range of 10^4 to 10^6 Hz, since in this window most of the signal power is available. Observe that also in the window $[10^6, 5 \times 10^6]$ there is an insignificant amount of signal power.

Table 3.2 Relative Signal Power Distribution Over a Selection of Frequency Bins

Frequency Bin	*Relative Power*
$[0, 500]$	0.0059
$[500, 5 \times 10^3]$	0.0257
$[5 \times 10^3, 10^4]$	0.0497
$[10^4, 10^5]$	0.2067
$[10^5, 10^6]$	0.6298
$[10^6, 5 \times 10^6]$	0.0822
$[10^2, 10^6]$	0.9998
$[10^3, 10^6]$	0.9942
$[10^4, 10^6]$	0.9228
$[0, 10^3]$	0.0059
$[0, 10^4]$	0.0669
$[0, 10^5]$	0.2839
$[0, 10^6]$	1

3.4.2 Power Density for Wideband and Ultrawideband Signals

The example considered in the Section 3.4.1 illustrated the fact that it will be somewhat difficult to populate (3.13) as time evolves. However, it is possible to propose more practical solutions based upon the observations in the previous section. Recall that the given pulse has a very short width in time, and during this time interval the full spectrum of frequencies that the waveform admits will be emitted. The example of the previous section illustrated that it is possible to partition the frequency range into intervals where the signal power in this interval can be estimated. Hence one can suppose that this signal power is uniformly distributed across this frequency bin. Therefore suppose that the frequency range is partitioned into disjoint sets $\{A_j, j \in J\}$ and that the power distribution is given by $\{P_j, j \in J\}$. Then one may assume that for any $f_j \in A_j$ its associated power is $P_j/|A_j|$, where $|A_j|$ is the size of the set A_j. Since the pulse duration will be assumed very short one may then produce a sequence of power density functions $\{I_j(t), j \in J\}$ where in view of (3.8) this power and the wavelength corresponding to the frequency can be applied. The term A_R will of course depend on the associated wavelength, while A_T and $R(t)$ will remain the same. It will thus be assumed that the HPRF DEW will be transmitting the pulse train continuously until the desired effect will be achieved.

One of the consequences of this approach is that if the frequency bin is sufficiently wide then the average power will be reduced to an insignificant proportion. This will result in performance prediction results, through the sequence of power density functions, indicating that the DEW will be completely ineffective, unless sufficiently high power levels are available. Therefore it is important to propose an alternative. A conclusion from the example in the Section 3.4.1 is that most of the signal power is concentrated in the frequency bin of $[10^4, 10^6]$ Hz, and in particular 922,800W of power is concentrated in this window. One can make the assumption that the power is distributed across the three frequencies of 10^4, 10^5, and 10^6 Hz; it remains to determine how to proportion the total power in this window across these three frequencies. One method is to determine, from the power spectral density, weightings to be applied to determine the proportions. To explain this, the vector of 256 frequencies were scanned to determine which almost matched these three frequencies. It was found that the second sampled frequency was $f_a = 1.9531 \times 10^4$ Hz, which is the nearest to 10^4. The power spectral density at this frequency is $w_1 = 7.0004$ watts per Hz (W/Hz). The sixth frequency is $f_b = 9.7656 \times 10^4$, which is the nearest to 10^5 Hz, with a power spectral density of $w_s = 232.2342$ W/Hz. Finally, the fifty-second frequency is $f_c = 9.9609 \times 10^5$, which is the nearest to 10^6, with a power spectral density of $w_3 = 0.0907$ W/Hz. Hence one may apportion the total power through weightings constructed from the power spectral density. Hence for f_b, one can assign the power of $w_1/(w_1 + w_2 + w_3) \times 10^6$, producing a power of $P_a = 2.9251 \times 10^4$ W corresponding to frequency f_a. Similarly, for f_b the corresponding power is $P_b = 9.7037 \times 10^5$ W and for f_c it is $P_c = 379.0261$W. With these results one may construct three power density functions of the form (3.8), and in Chapter 4 it will be shown how these can be combined to produce a single measure for approximate performance prediction of an HPRF DEW operating a continuous pulse train of signals of the form (3.14).

It is important to note that if the sampling rate of the signal is sufficiently large then one may produce a better distribution of powers in a wider frequency range. Another important factor worth considering here is that the waveform has focused the frequency distribution into the range $[10^4, 10^6]$, which will correspond to considerable wavelengths. Coupling is possible in this frequency range but the distribution of power may limit the utility of a deployed HPRF DEW using this waveform characteristic. The concentration of frequencies, as illustrated in Figures 3.7 and 3.8, is an attribute of the signal (3.14). By examining Gaussian modulated signals one may produce a better distribution of frequencies for HPRF DEW effects, as discussed in [13].

3.5 Concluding Remarks

This chapter's primary focus has been to demonstrate how a power density function can be produced to measure the power that a HPRF DE system focuses on a target. Based upon this it was shown how its statistical nature could be used to produce probability measures of likely power on a threat. A series of benchmark estimates of power levels required to produce desired disruptive effects, appropriate frequencies, and system-level target vulnerability thresholds were also documented, based upon results derived from experimental campaigns reported in the open literature. Both the narrowband and wideband cases were examined and a number of results were produced to demonstrate the relevant results. These will be applied in Chapter 4 to demonstrate the quantification of HPRF DE system performance.

References

[1] Benford, J., and J. Swegle, *High-Power Microwave*, Norwood, MA: Artech House, 1992.

[2] Taylor, C. D., and D. V. Giri, *High-Power Microwave Systems and Effects*, Washington, D.C.: Taylor and Francis, 1994.

[3] Giri, D., *High-Power Electromagnetic Radiators: Nonlethal Weapons and Other Applications*, Cambridge, MA: Harvard University Press, 2004.

[4] Giri, D. V., R. Hoad, and F. Sabath, *High-Power Electromagnetic Effects on Electronic Systems*, Norwood, MA: Artech House, 2020.

[5] Cristina, S., and A. Orlandi, "Calculation of the Induced Effects due to a Lightning Stroke," *IEE Proceedings-Part B*, Vol. 139, 1992, pp. 47–55.

[6] Orlandi, A., "Lightning Induced Transient Voltages in Presence of Complex Structures and Nonlinear Loads," *IEEE Transactions on Electromagnetic Compatibility*, Vol. 38, No. 2, 1996, pp. 150–155.

[7] Perez, R. J., *Handbook of Aerospace Electromagnetic Compatibility*, New Jersey: Wiley-IEEE Press, 2019.

[8] Radasky, W. A., C. E. Baum, and W. M. Wik, "Introduction to the Special Issue on High Power Electromagnetics (HPEM) and Intentional Electromagnetic Interference (IEMI)," *IEEE Transactions on Electromagnetic Compatibility*, Vol. 46, 2004, pp. 314–321.

[9] Giri, D. V., and F. M. Tesche, "Classification of Intentional Electromagnetic Environments," *IEEE Transactions on Electromagnetic Compatibility*, Vol. 46, 2004, pp. 322–328.

[10] Kovalev, N. F., "Cherenkov Microwave Generation of High-Current Relativistic Electron Beams," in *IEEE 9th International Conference on High-Power Particle Beams*, Vol. 3, 1992.

[11] Gai, W., "High-Power RF Generation by Relativistic Beams in Dielectric Structures and its Application to the Advanced Accelerators," in *IEEE 25th International Conference on Infrared and Millimeter Waves*, 2000.

[12] Okada, T., and K. Niu, "Electromagnetic Instability and Stopping Power of Plasma for Relativistic Electron Beams," in *IEEE 3rd International Topical Conference on High-Power Electron and Ion Beam Research and Technology*, 1979.

[13] Hong, K. D., and S. W. Braidwood, "Resonant Antenna-Source System for Generation of High-Power Wideband Pulses," *IEEE Transactions on Plasma Science*, Vol. 30, No. 5, 2002, pp. 1705–1711.

[14] Sarkar, P., et al., "A Compact Battery-Powered 500 kV Pulse Generator for UWB Radiation," in *IEEE Pulsed Power Conference*, 2005.

[15] Giri, D. V., and F. M. Tesche, "High Power Electromagnetic (HPEM) Source Considerations," in *IEEE International Symposium on Electromagnetic Compatibility*, 2003, pp. 1028–1031.

[16] Giri, D. V., R. Hoad, and F. Sabath, "Implications of High-Power Electromagnetic (HPEM) Environments on Electronics," *IEEE Electromagnetic Compatibility Magazine*, Vol. 9, No. 2, 2020, pp. 37–43.

[17] Radasky, W. A., "The Threat of Intentional Interference (IEMI) to Wired and Wireless Systems," in *17th International Zurich Symposium on Electromagnetic Compatibility*, 2006.

[18] Hoad, R., A. Lambourne, and A. Wraight, "HPEM and HEMP Susceptibility Assessments of Computer Equipment," in *17th International Zurich Symposium on Electromagnetic Compatibility*, 2006.

[19] Clarke, T., et al., "Predictive Modeling of High-Power Electromagnetic Effects on Electronics," in *International Conference on Electromagnetics in Advanced Applications*, 2011.

[20] Bayram, Y., et al., "High Power EMI on RF Amplifier and Digital Modulation Schemes," *IEEE Transactions on Electromagnetic Compatibility*, Vol. 50, 2008, pp. 849–860.

[21] Kasmi, C., and J. Lopes-Esteves, "Functional Susceptibility of COTS Devices to IEMI at Local and Large-Scale Levels," in *International Conference on Electromagnetics in Advanced Applications*, 2016.

[22] Esteves, J., E. Cottais, and C. Kasmi, "Unlocking the Access to the Effects Induced by IEMI on a Civilian UAV," in *International Symposium on Electromagnetic Compatibility*, 2018, pp. 48–52.

[23] Graswald, M., et al., "Defeating UAVs Through Novel HPEM Effectors," in *31st International Symposium on Ballistics, India*, 2019.

[24] He, K., et al., "Detection and Identification of System-Level Soft Failure Induced by Radio Frequency Interference in Small UAV System," *IEEE Transactions on Electromagnetic Compatibility*, Vol. 64, No. 3, 2022, pp. 661–673.
[25] Zhao, M., et al., "Investigation on Falling and Damage Mechanisms of UAV Illuminated by HPM Pulses," *IEEE Transactions on Electromagnetic Compatibility*, Vol. 64, No. 5, 2022, pp. 1412–1422.
[26] Esteves, J., E. Cottais, and C. Kasmi, "Analysis of HPEM Perturbations Induced on the Navigation System of a UAV," in *AMEREM*, 2018.
[27] Lubkowski, G., et al., "Response of the UAV Sensor System to HPEM Attacks," in *International Symposium on Electromagnetic Compatibility*, 2020.
[28] Lavau, L. C., M. Suhrke, and P. Knott, "Susceptibility of Sensors to IEMI Attacks," in *IEEE International Joint EMC/SI/PI and EMC Symposium*, 2021, pp. 533–537.
[29] Sakharov, K., et al., "Study of UWB Electromagnetic Pulse Impact on Commercial Unmanned Aerial Vehicle," in *International Symposium on Electromagnetic Compatibility*, 2018, pp. 40–43.
[30] Esteves, J. L., "Electromagnetic Watermarking: Exploiting IEMI Effects for Forensic Tracking of UAVs," in *International Symposium on Electromagnetic Compatibility*, 2019, pp. 1144–1149.
[31] Braidwood, S., and K. Hong, "Stopping Car Engines using High-Power Electromagnetic Pulses," in *International Conference on Electromagnetics in Advanced Applications*, 2010.
[32] Cai, J., Z. Wang, and F. Qin, "Wideband High-Power Microwave Coupling Analysis of Automotive Engine," in *IEEE International Conference on Numerical Electromagnetic Modelling and Optimization for RF, Microwave and Terahertz Applications (NEMO)*, 2020, pp. 1–3.
[33] Stimson, G. W., *Introduction to Airborne Radar*, Raleigh, NC: Scitech, 1998.
[34] Weinberg, G. V., "Quantification of Combat Team Survivability with High-Power RF Directed Energy Weapons," *Progress in Electromagnetics Research M*, Vol. 102, 2021, pp. 1–11.
[35] Friis, H. T., "A Note on a Simple Transmission Formula," *IRE Proceedings*, Vol. 34, No. 5, 1946, pp. 254–256.
[36] Levanon, N., *Radar Principles*, New York: John Wiley & Sons, 1988.
[37] Mokole, E. L., M. Kragalott, and K. R. Gerlach, "*Ultra-Wideband, Short Pulse Electromagnetics 6*," Boston, MA: Springer Science and Business Media, 2003.
[38] Ghavami, M., L. B. Michael, and R. Kohno, *Ultra Wideband Signals and Systems in Communications Engineering*, New York: John Wiley & Sons, 2007.

CHAPTER

4

Contents

4.1 One Threat

4.2 Multiple Threats

4.3 Concluding Remarks

References

HPRF Performance Prediction

The purpose of this chapter is to apply the results of the previous one to demonstrate how HPRF DEW performance prediction can be undertaken. Throughout this chapter it will be supposed that a single HPRF DEW is providing defense against threats. Initially a single threat will be considered, and this will then be extended to consider groups of threats. In terms of target characteristics the power density functions derived in the previous chapter depends on both the distance from the effector to the threat as well as the threat's coupling cross section. A power density function over an observational window requires the specification of the time when a threat arrives. Hence this can be either assumed to be instantaneous, such as the case where a swarm of UAVs arrive into the HPRF DEW's beam or they may be assumed to arrive via a stochastic process. In addition to this, the manner in which threats arrive can determine an appropriate model for processing their defeat. Hence a swarm of UAVs may be distributed spatially, while a series of missiles fired toward the HPRF DEW may be assumed to travel in a linear path. Consequently in the former scenario one can examine the defeat in terms of coverage area of the HPRF DEW's main beam, while in the latter it may be necessary to model the process through a single server queue with service time delays.

4.1 One Threat

This section provides some examples of the ways in which performance prediction can be performed for a single threat being countered by a solo HPRF DEW. The case of a narrowband source will be shown to be somewhat straightforward, while the extension to a wideband or ultrawideband effector is more complex, but simplified by the analysis at the end of the previous chapter.

4.1.1 Narrowband Case

In the situation where a single HPRF DEW is radiating a single target, performance prediction can be based upon (3.8) with appropriate assumptions adopted for A_R from a statistical perspective. Suppose that $\tau > 0$ is the target vulnerability threshold, in W/m^2, such that when the power density $I(t)$ on the target exceeds it a system effect is produced. Hence the probability that the target is disrupted by time t is

$$\mathbf{P}(\text{Threat disrupted by time } t) = \mathbf{P}(I(t) > \tau). \qquad (4.1)$$

In the first instance suppose that τ is deterministic and that A_R is modeled as an exponential distribution with parameter μ. Then an application of (3.10) yields

$$\mathbf{P}(\text{Threat disrupted by time } t) = e^{-\frac{\mu\tau\lambda^2 R^2(t)}{P_T A_T}}, \qquad (4.2)$$

which can be used to produce performance prediction results. To illustrate this, consider the case of an HPRF DEW operating at a frequency of 1 GHz with an antenna with effective area $A_T = 1$ m^2, being subjected to a threat on a direct course for the DE system. It is assumed that this threat travels on a straight path at a speed of 40 m/s, and will take 240s to reach the DEW's location if it is not disrupted. Hence $R(t) = 40(240 - t)$ for $0 \leq t \leq 240$, and this threat originates at a distance of 9.6 km from the DEW. This threat's coupling cross section is modeled through an exponential distribution with parameter $\mu = 0.1$ and its coupling threshold is assumed to be $\tau = 10$. Figure 4.1 plots (4.1) for the four DEW power levels of $P_T = 10^4$, 10^5, 10^6, and 10^9 W, respectively. From the

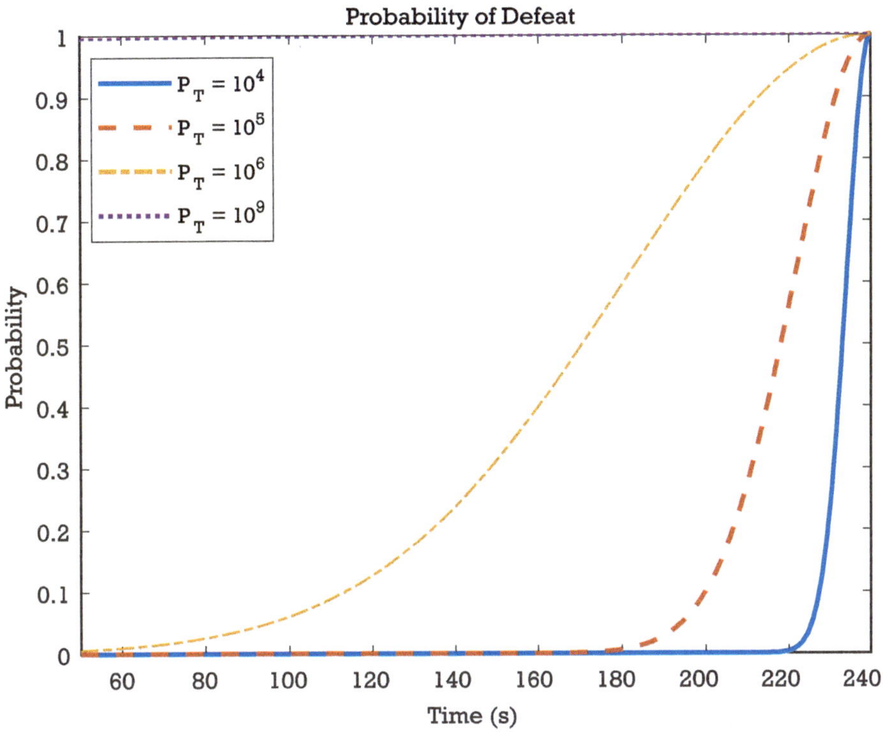

Figure 4.1 Four examples of HPRF DEW performance at a frequency of 1 GHz, with a threat traveling at 40 m/s.

figure, one observes that the greater the power the more likely the threat is disabled earlier; in particular, one can see that with a GW of power the target is almost certainly disabled once it arrives into the DEW's beam.

One important observation concerning Figure 4.1 is the fast rise toward probability unity as time is near the upper boundary of the travel time of the threat. As the speed of the threat is increased, convergence to probability unity is through a steeper function. Figure 4.2 demonstrates this effect; this figure repeats the scenario of Figure 4.1 except the threat is traveling at 300 m/s, and it is assumed that it takes the same 240s period to reach the DEW if not disrupted. Observe that the DEW operating with a peak power of 10^4 W will have to defeat this threat in a time window of length 5s before it would reach the DEW. Although this probability still reaches unity, the concern is

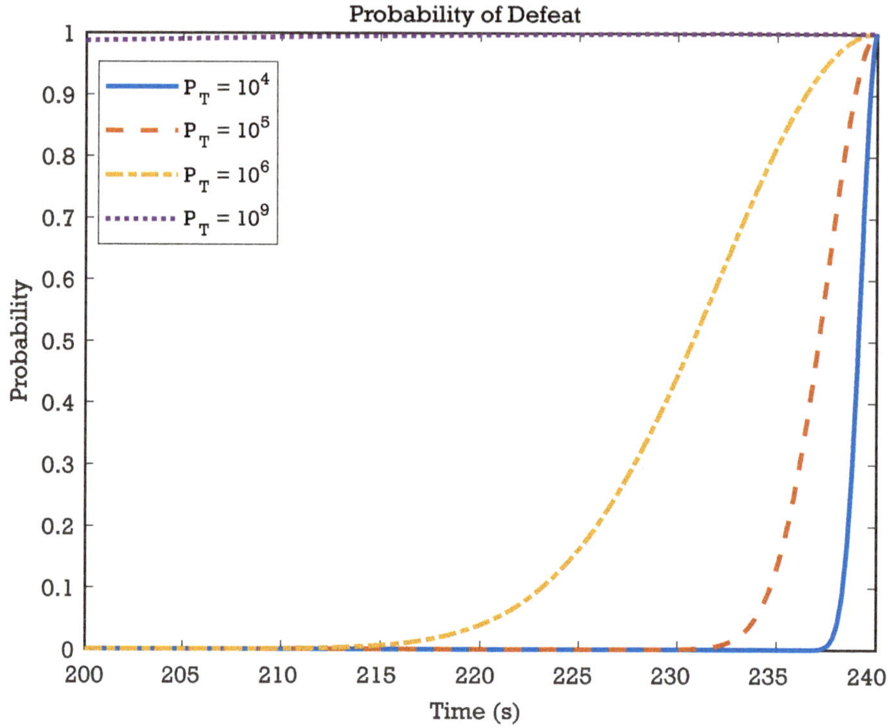

Figure 4.2 A repeat of the scenario of Figure 4.1, except the threat is traveling at 300 m/s.

that there may be insufficient stand-off time available in practice to defeat the threat. Hence performance prediction curves such as those in Figures 4.1 and 4.2 must be interpreted very carefully. The DEW with a power level of 1 GW is still likely to defeat the threat because it can deliver sufficient power over a considerable range. These concerns with interpretation of performance curves, in the context of HEL DEWs, have been discussed in [1].

While a continuously radiating HPRF effector can transfer energy onto the target once it is very close to the DEW platform, this rapid convergence to unity can result in misunderstandings of performance.

One may in fact suppose that the threshold τ in (4.1) is not deterministic but modeled through a random variable. In particular, assume that τ is a continuous random variable with support S and probability density function

$f_\tau(x)$. Then through an application of the result (2.18) one can show that (4.1) becomes

$$\mathbf{P}(\text{Threat disrupted by time } t) = \int_S f_\tau(x)\mathbf{P}(I(t) > x \mid \tau = x)dx, \tag{4.3}$$

and it may be assumed that the threshold and A_R are independent, so that the conditional probability in (4.3) may be eliminated.

Suppose that A_R has the same exponential distribution adopted previously, and that τ has a uniform distribution on the interval $[\tau_0, \tau_1]$. Then applying the density (2.10) to (4.3) and evaluating the integrals results in

$$\begin{aligned}\mathbf{P}(\text{Threat disrupted by time } t) = {} & \frac{1}{\tau_1 - \tau_0}\frac{P_T A_T}{\mu\lambda^2 R^2(t)} \\ & \times \left[e^{-\tau_0 \frac{\mu\lambda^2 R^2(t)}{P_T A_T}} - e^{-\tau_1 \frac{\mu\lambda^2 R^2(t)}{P_T A_T}} \right]. \end{aligned} \tag{4.4}$$

To illustrate the application of (4.4) it is applied to the example utilized in Figure 4.1, except the threshold is assumed to be uniformly distributed over the interval [10, 100]. Figure 4.3 shows predicted performance, and when compared with the results in Figure 4.1 one can observe the decrease in performance due to the increase in uncertainty about the disruption threshold.

4.1.2 Wideband and Ultrawideband Case

As described in the context of the example in Section 3.4.1 it is supposed that a series of power density functions have been produced, namely $\{I_j(t), j \in J\}$, where each $I_j(t)$ is of the form given by (3.8). Specifically,

$$I_j(t) = \frac{P_{Tj} A_T A_{Rj}}{\lambda_j^2 R(t)^2}, \tag{4.5}$$

where at each frequency f_j corresponding to wavelength λ_j there is P_{Tj} watts of power available, and the coupling cross section of the target is also now assumed to vary with frequency, as indicated through A_{Rj}. It is supposed that the pulse width is sufficiently short so that the discussions in the previous chapter apply. The antenna is assumed to have a common effective area A_T and the distance to the target is still $R(t)$. It is assumed that for each frequency there is a threshold τ_j such that if the power density function on the target exceeds this threshold then coupling has occurred at the corresponding frequency.

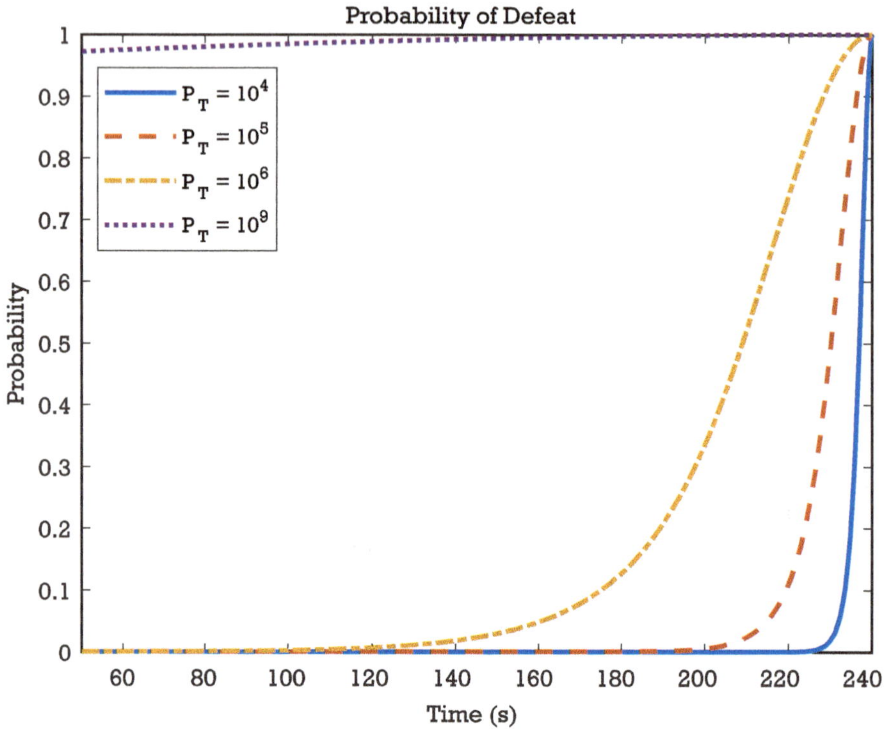

Figure 4.3 A repeat of the scenario in Figure 4.1, except the threshold is assumed uniformly distributed in the interval [10, 100].

The probability that the threat is defeated by time t is the probability that at least one of the power density functions exceeds the corresponding threshold. Hence assuming these occur independently,

$$\begin{aligned}\mathbf{P}(\text{Threat disrupted by time } t) &= \mathbf{P}(\text{At least one } I_j(t) > \tau_j) \\ &= 1 - \mathbf{P}(\text{None of the } I_j(t) > \tau_j) \\ &= 1 - \prod_{j \in J} \mathbf{P}(I_j(t) \leq \tau_j). \qquad (4.6)\end{aligned}$$

Therefore, (4.6) provides a framework in which the performance prediction of a wideband or ultrawideband HPRF DE system may investigated. It is now informative to examine a particular example. As in the derivation of (3.10) suppose that the coupling cross section is modeled through an exponential distribution; specifically suppose that A_{Tj} has an exponential distribution with parameter μ_j. Then (4.6) reduces to

$$\mathbf{P}(\text{Threat disrupted by time } t) = 1 - \prod_{j \in J} \left(1 - e^{-\frac{\mu_j \tau_j \lambda_j^2 R^2(t)}{P_{Tj} A_T}} \right). \tag{4.7}$$

The example of a damped sinusoidal signal from the previous chapter will be applied here for illustration purposes. Hence suppose that the HPRF DEW operates at the three frequencies of 1.9531×10^4, 9.7656×10^4, and 9.9609×10^5 Hz, respectively, with power levels of 2.9251×10^4, 9.7037×10^5, and 379.0261W. Due to the corresponding large wavelengths the choice of $A_T = 10^5$ has been made. As in two of the previous examples it is assumed that the threat is traveling at a speed of 40 m/s and will take 240s to reach the DEW's location if the threat has not been disrupted. Figure 4.4 shows four

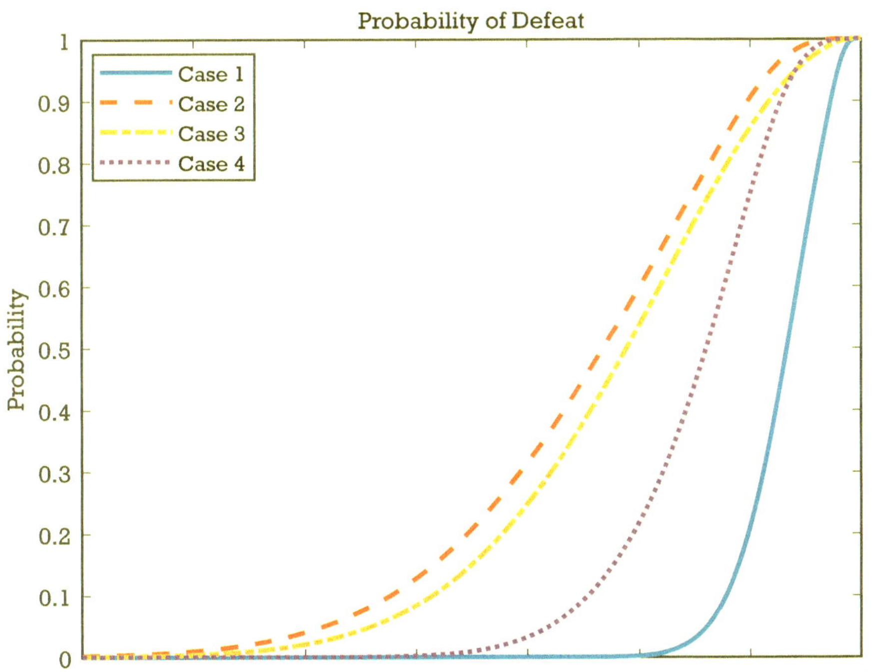

Figure 4.4 Four cases to illustrate the predicted performance of an ultrawideband HPRF DEW bases upon (4.7).

cases of (4.7). In all cases the thresholds are assumed to be unity. Case 1 is for the situation where each $\mu_j = 0.1$; Case 2 changes μ_1 to 10^{-5} while preserving the other target parameters. Case 3 applies $\mu_2 = 0.01$ while the other target parameters are set to 0.1. Finally, Case 4 sets $\mu_3 = 10^{-3}$ while the other target parameters are set again to 0.1. This example demonstrates the effect of the reduction in DEW power, since threat defeat is not likely until the threat has been in transit well after the 180-second mark. In Case 1 the worst overall performance is observed. This improves only as the target parameter μ_j is decreased, which corresponds to an increase in the target's mean parameter. As one of the target parameters is changed to smaller values, while preserving the others at 0.1, one observes a change in overall performance. Case 2 corresponds to permitting the threat to have the greatest visibility in the second frequency example, which also corresponds to the greatest power level, and so the gain in disruption likelihood is not unexpected.

4.2 Multiple Threats

In more realistic scenarios one can expect the HPRF DEW to be presented with a series of threats, and so it is important to extend the previous section's development to this case. However, throughout the following a restriction will be imposed that the HPRF DEW is narrowband, and so the power density function (3.8) will be assumed. One important factor to now be introduced is the modeling of threat arrival times. In the first case it will be supposed that the threats arrive into the DEW's operational space at the same time. Subsequently, it will be shown how their arrival times will be modeled through a renewal process, as introduced in Chapter 2. The results presented in this section and the following one have appeared in [2].

4.2.1 Threats Arrive Simultaneously

Suppose that all threats arrive into the HPRF DEW's theater of operation initially, and suppose that there are a fixed number N of such threats. Then the power density function of the HPRF DEW on threat j over the observational window $[0, t]$ is given by

$$I_j(t) = \int_0^t \frac{P_T A_T A_R}{\lambda^2 R_j(s)^2} ds, \tag{4.8}$$

where $R_j(t)$ is the distance between the HPRF DEW and threat j at time t.

As in the single threat case, the required condition for the defeat of threat j is that (4.8) exceeds a disruption threshold τ. Throughout the following the threats will be assumed to be homogeneous. Otherwise one can index the threshold by j to indicate it varies with different threats. This threshold will be considered to be a system-level disruption threshold, and is also measured in units of W/m^2. As mentioned in the previous chapter, A_R can be modeled through an exponential distribution with parameter μ_T.

The probability of defeating threat j by time t is given by the likelihood that (4.8) exceeds τ. Defining this probability to be $\phi_j(t)$ it therefore follows that by using the approach in Chapter 3 that

$$\phi_j(t) = e^{-\frac{\mu_T \tau \lambda^2}{P_T A_T}\left[\int_0^t R_j^{-2}(s)ds\right]^{-1}}, \tag{4.9}$$

for each $1 \leq j \leq N$. It is now useful to count the number of threats defeated by time t, since (4.9) provides an appropriate success probability, in the terminology of Bernoulli trials [3]. Hence define $M(t)$ to be the number of threats defeated by time t. Then as a sum of Bernoulli events one can write

$$M(t) = \sum_{j=1}^{N} M_j(t) \tag{4.10}$$

where for each j

$$M_j(t) = \begin{cases} 1 & \text{with probability } \phi_j(t) \\ 0 & \text{with probability } 1 - \phi_j(t), \end{cases} \tag{4.11}$$

is representative of each Bernoulli trial with success probability given by (4.9).

One can use (4.10) to assess performance of the HPRF DEW, since the sequence of variables defined through (4.11) are independent, and consequently (4.10) is a generalized binomial distribution. However, it is more practical to consider the mean of (4.10) to construct a performance measure. Consequently the mean of (4.10) is

$$\begin{aligned} \mathbf{E}(M(t)) &= \sum_{j=1}^{N} \phi_j(t) \\ &= \sum_{j=1}^{N} e^{-\frac{\mu_T \tau \lambda^2}{P_T A_T}\left[\int_0^t R_j^{-2}(s)ds\right]^{-1}}. \end{aligned} \tag{4.12}$$

Therefore (4.12) provides a metric for the performance prediction of an HPRF DEW in the situation where a series of threats have appeared into its theater of operation at the same time.

It is now useful to specialize this to particular examples. To illustrate the application of (4.12), assume that the series of threats are converging on the HPRF DEW, such that at time t $R_j(t) = \nu_j(T_j - t)$, for $t \leq T_j$. Hence all the threats are traveling toward the HPRF DE source at a constant speed of ν_j m/s, and will take T_j seconds to arrive at the effector's location if not disrupted. It is important to note that the threats may be approaching from different directions toward the DEW, as illustrated in Figure 4.5. If the threats are in a linear path to the HPRF DEW then one must model the situation differently, as will be discussed toward the end of this chapter.

Under the modeling assumptions imposed, it can be shown that (4.12) reduces to

$$\mathbf{E}(M(t)) = \sum_{j=1}^{N} e^{-\frac{\mu_T \tau \lambda^2}{P_T A_T} \frac{\nu_j (T_j - t) T_j}{t}}, \tag{4.13}$$

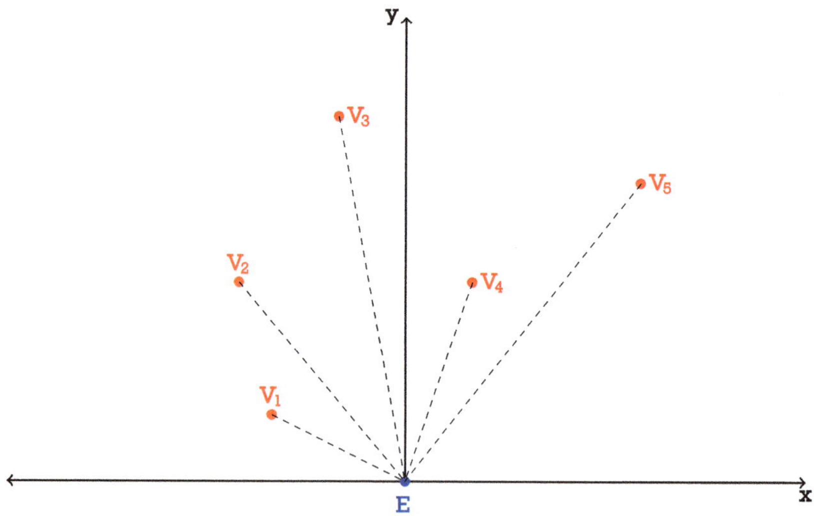

Figure 4.5 A series of five threats converging onto the HPRF DEW. These threats have been supposed to arrive into the DEW's theater of operation simultaneously, but have appeared at different places from a geometric position, traveling at different speeds and taking different times to arrive to the DEW's location if not disrupted.

provided $t < \min_{j\in\{1,2,\ldots,N\}} T_j$. Therefore (4.13) can be applied for performance prediction for a single HPRF DEW countering a series of N threats that arrive simultaneously.

Two examples will now be considered to illustrate (4.13) from a numerical perspective. Suppose that there are $N = 100$ threats, each traveling at 40 m/s and such that each would take 120s to reach the HPRF source if not disrupted. It is supposed that these homogeneous threats have $\mu = 10$, so that they have a small coupling cross section. The DEW operates with a power level of 1 GW at a frequency of 1 GHz, with $A_T = 1$ m^2. Figure 4.6 shows the mean number of threats defeated as a function of t, for four vulnerability thresholds: 10^2, 10^3, 10^4, and 10^5 W/m^2. The figure shows that as the vulnerability threshold increases it will take substantially longer to defeat the swarm of threats, which is not a surprising result.

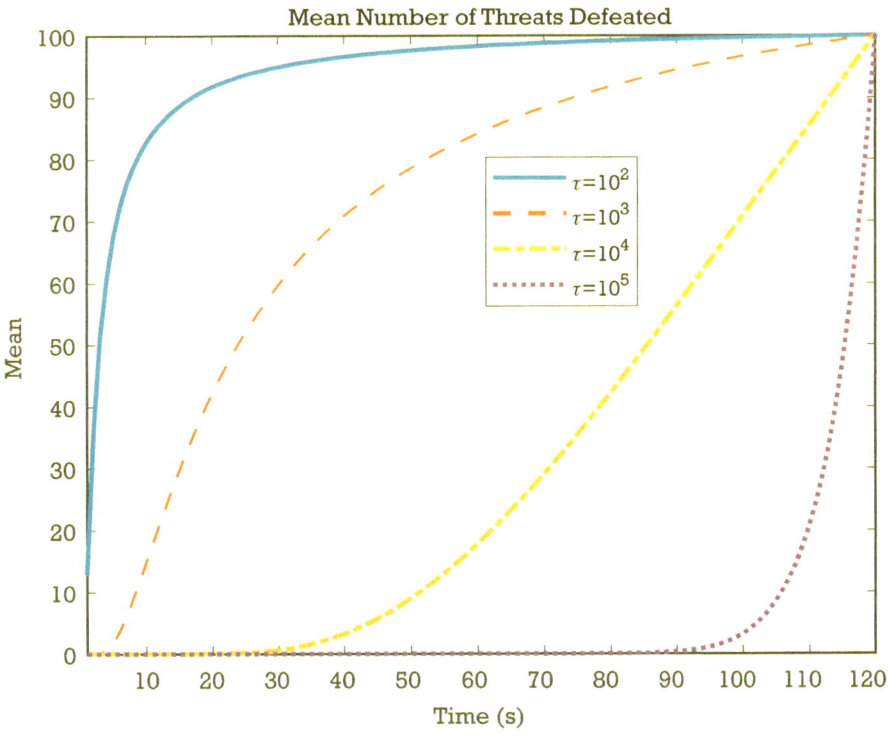

Figure 4.6 An illustration of (4.13) for four vulnerability thresholds.

Figure 4.7 shows the same scenario except that the threat coupling cross section parameter is now 0.1, which implies that there is a greater coupling cross section. As can be observed from the figure, for smaller vulnerability thresholds there is sufficient power to disable the threats once they arrive into the DEW's beam. For larger vulnerability thresholds it will likely take longer to disable the complete swarm.

4.2.2 Arrivals Through a Renewal Process

It is somewhat more realistic to assume that the threats arrive randomly into the HPRF DEW's beam, from the point of view of the HPRF DE effector. Hence this section introduces the concept of modeling arrivals through a renewal process, as introduced in Chapter 2. Hence it will be supposed that threats arrive according to a renewal process $\{S_j, j \in \mathbb{N}\}$ where S_j is the

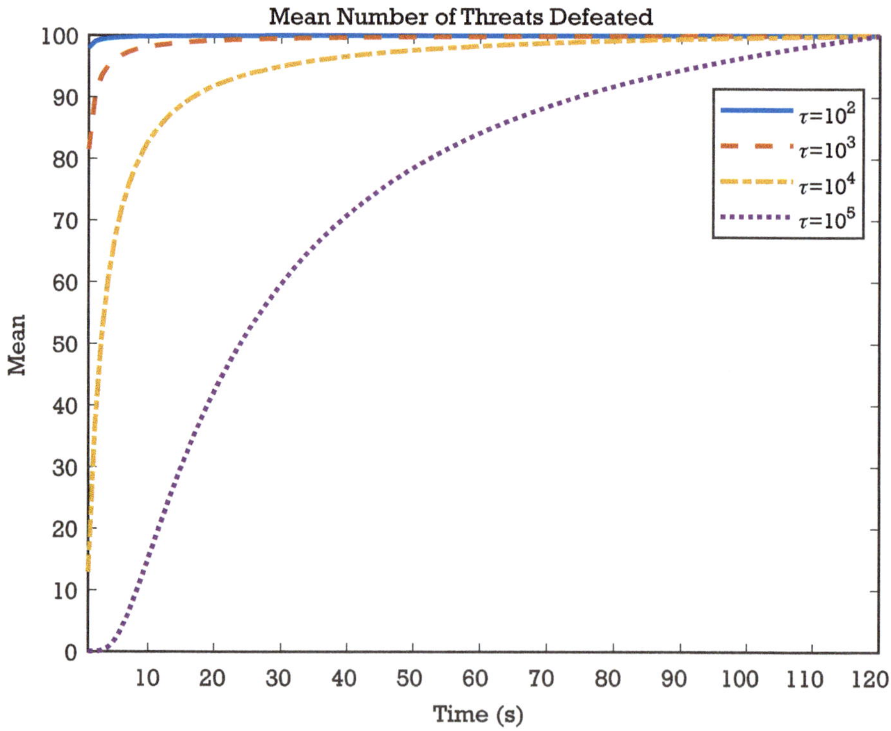

Figure 4.7 A second illustration of (4.13) for the same four vulnerability thresholds but with $\mu = 0.1$.

arrival time of the jth threat. This implies that the threat interarrival times are independent and identically distributed. Consequently if $\{X_j, j \in \mathbb{N}\}$ is the interarrival process then $S_j = X_1 + X_2 + \cdots + X_j$.

When threat j arrives it is assumed that its arrival is into the effective area of the HPRF DEW, and thus it is radiated from its arrival time, as in the Section 4.2.1. Therefore if t is the amount of time the jth threat has been radiated after its arrival time, then given a specific S_j the power density on this threat over the observational window $[S_j, S_j + t]$ is given by

$$\int_{S_j}^{S_j+t} I_j(w)dw, \tag{4.14}$$

where the intensity $I_j(w)$ is given by (3.8).

What is required is determination of each threat's service time relative to its arrival process. Hence suppose that Y_j is the jth threat's service time after arrival, or equivalently the time after arrival that it takes for the HPRF effector to cause a disruption. Then conditioned on S_j, the probability that Y_j exceeds t is exactly the probability that (4.14) is smaller than the disruption threshold τ, understood to also be conditional on S_j. Therefore

$$\mathbf{P}(Y_j > t | S_j) = \mathbf{P}\left(\int_{S_j}^{S_j+t} I_j(w)dw < \tau\right). \tag{4.15}$$

Under the assumption that the random variable A_R specified in the definition of (3.8) is independent of S_j it can be shown that

$$\mathbf{P}(Y_j > t | S_j) = 1 - e^{-\frac{\mu_T \tau \lambda^2}{P_T A_T}\left[\int_0^t R_j^{-2}(w+S_j)dw\right]^{-1}}, \tag{4.16}$$

which has been derived using a similar analysis as in the derivation of (4.9). Noting that the probability of threat j being defeated by time t is equivalent to the event that $S_j + Y_j < t$ it follows that

$$\begin{aligned}\mathbf{P}(S_j + Y_j < t) &= \int_0^{\infty} f_{S_j}(s)\mathbf{P}(S_j + Y_j < t | S_j = s)ds \\ &= \int_0^t f_{S_j}(s)\mathbf{P}(Y_j < t - s | S_j = s)ds, \end{aligned} \tag{4.17}$$

which has been obtained through conditional probability, noting that the random variable Y_j is nonnegative, and where $f_{S_j}(w)$ is the density of S_j.

Hence an application of (4.16) to (4.17) yields

$$\mathbf{P}(S_j + Y_j < t) = \int_0^t f_{S_j}(s) e^{-\frac{\mu\tau\lambda^2}{P_T A_T}\left[\int_0^t R^{-2}(w+S_j)dw\right]^{-1}} ds. \tag{4.18}$$

This expression thus permits one to determine the statistical nature of a threat's resultant lifetime.

Given a series of threats it is of interest to incorporate (4.18) into a measure for the number of threats that the HPRF DEW has defeated by a given time. Hence, define $\psi_j(t) = \mathbf{P}(S_j + Y_j < t)$, and as in the definition (4.11) let $M_j(t)$ be a random variable defined through

$$M_j(t) = \begin{cases} 1 & \text{if } S_j + Y_j < t \\ 0 & \text{otherwise.} \end{cases} \tag{4.19}$$

Then if there have been $N(t)$ arrivals through the renewal process by time t then

$$M(t) = \sum_{j=1}^{N(t)} M_j(t) \tag{4.20}$$

is a compound process counting the number of arrivals that have been defeated by time t. It will be difficult to work with (4.20) in practice, so as before its mean is examined for performance assessment. Hence by applying a double expectation

$$\mathbf{E}(M(t)) = \mathbf{E}(\mathbf{E}(M(t)|N(t))) = \mathbf{E}\left(\sum_{j=1}^{N(t)} \psi_j(t)\right). \tag{4.21}$$

Through an analysis of the evaluation of (4.21) in terms of point probabilities of $N(t)$ it can be shown that

$$\mathbf{E}(M(t)) = \sum_{j=1}^{\infty} \psi_j(t)\mathbf{P}(N(t) \geq j). \tag{4.22}$$

It is also interesting to observe that from renewal theory the events $\{N(t) \geq j\}$ and $\{S_j \leq t\}$ are equivalent and so

$$\mathbf{E}(M(t)) = \sum_{j=1}^{\infty} \psi_j(t)\mathbf{P}(S_j \leq t). \tag{4.23}$$

In order to provide illustrations of performance it will be necessary to assume a particular realization for the maximum number of threats within a given time interval. This will imply that $N(t)$ in (4.20) will be set to a particular value to provide tangible results.

Suppose that threats are on a linear trajectory towards the HPRF DEW as in the previous section, and that these paths are not necessarily identical. Therefore suppose that for each $j \in \mathbf{N}$ that $R_j(t) = \nu_j(T_j - t)$ for $0 \leq t \leq T_j$. Then it can be shown that (4.18) reduces to

$$\psi_j(t) = \int_0^t fs_j(s) e^{-\frac{\mu_T \tau \lambda^2 \nu_j^2}{P_T A_T} \frac{(T_j - t)(T_j - s)}{(t-s)}} ds, \quad (4.24)$$

provided $t < T_j$. This can now be applied to (4.21) if it is assumed that $N(t)$ is fixed and known, or otherwise can be incorporated into (4.22) or (4.23) if uncertainty is to be preserved in terms of the number of arrivals. Evaluation of (4.24) can be achieved through numerical integration.

Specific examples will now be investigated, and it will be assumed that the number of arrivals is deterministic and set to 50. Arrivals will be through a Poisson process with rates varying from 1, 2, 3, and then 10. Each threat is homogeneous, traveling toward the HPRF DEW in a straight line, at a speed of 40 m/s, and it is assumed that it will take 240s for the threats to reach the DEW if not intercepted. Note that these straight arrival paths are not assumed identical. The HPRF DEW operates at a frequency of 1 GHz with 1 GW of power, and $A_T = 1$ m^2 as before. Suppose that each threat has a vulnerability parameter of $\mu = 10$ and that the corresponding disruption threshold is 1 kW per meter square. Then Figure 4.8 plots the mean number of threats defeated as the threat arrival rate is increased. As can be observed, as the arrival rate is increased the mean number of threats defeated can experience a decline, which is more significant as the rate reaches 10. Once the latter rate is achieved it is clear that the HPRF DEW will not disrupt all of the threats within the time window plotted. This is due to the fact that not all the threats may have arrived into the DEW's main beam for a time period sufficiently long to be illuminated adequately for a disruption to occur.

Suppose that the arrival rate is set to 2, while all the other parameters are maintained with the exception of the target vulnerability threshold. Figure 4.9 plots the resultant mean number of threats defeated for the four cases where this threshold is 10^3, 10^4, 10^5, and 10^6 W/m^2. One observes that as the threshold increases the resultant mean tends to decrease significantly

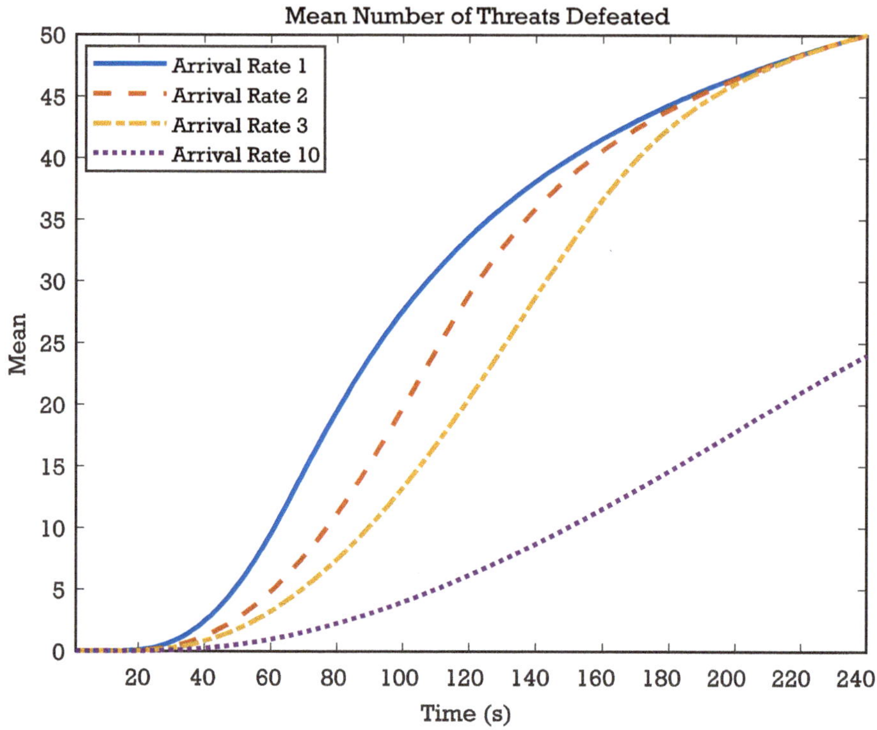

Figure 4.8 A series of plots showing the mean number of threats defeated over time, for four arrival rates.

over time. As the threshold reaches 10^6 W/m^2 it is clear that the behavior in the plot is as reported in [1] and must be interpreted carefully. Although the curve for this case spikes up to the maximum number of threats, this result suggests that all threats on average are defeated in a very small time window. Given the target vulnerability parameter is small in this case it may be indicative of the HPRF DEW reaching its performance limits. The solution, if it is required to be able to counter such threats with greater certainty, is to increase the DEW's overall power. To illustrate this, the same numerical experiment is performed except the DEW's power has been increased to 10^{12} W, and the results are plotted in Figure 4.10. As this figure demonstrates, the given system is in a better position to eliminate all the 50 threats on average within the given time interval. It is important to observe that

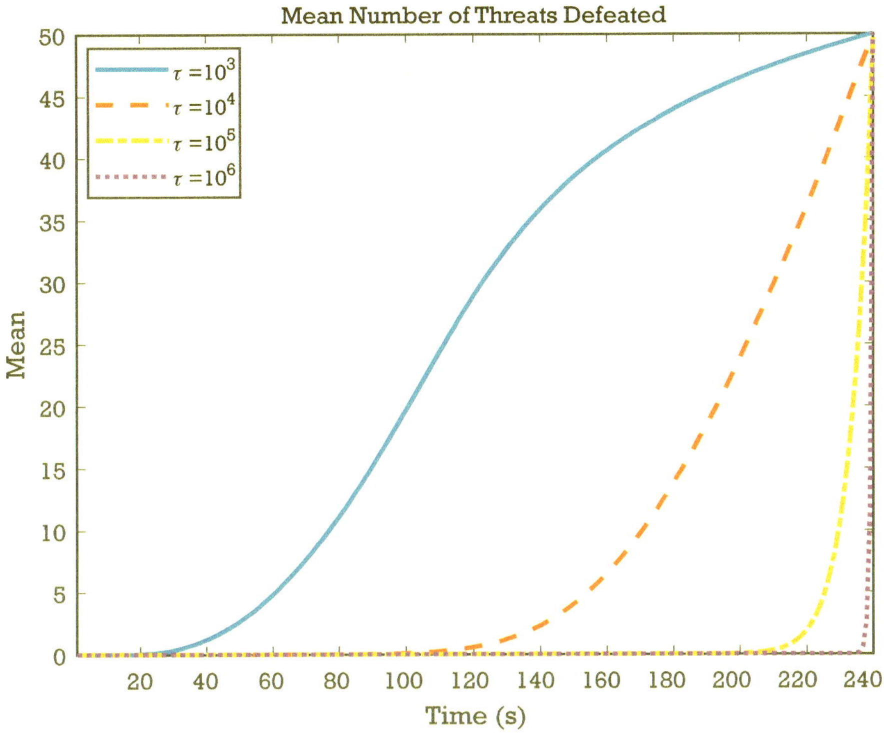

Figure 4.9 Plots of the mean number of threats defeated over time, for four vulnerability threshold levels.

increasing the power to this level and achieving such a result is under ideal situations and may not be a feasible solution in practice.

4.2.3 Threats Arrive Linearly

The underlying assumption in this case is that a sequence of threats have been directed toward the HPRF DEW position, such as a series of rocket-propelled grenades (RPGs), which travels toward the HPRF DEW in an approximate common straight line path. In the case of such threats the hope is that the HPRF DEW, through inducing currents on the surface of the RPG, may cause it to trigger early. UAVs may also be directed toward the HPRF source in a straight line; if they are sufficiently shielded then as time progresses it may be likely that the UAVs are not defeated until they are near

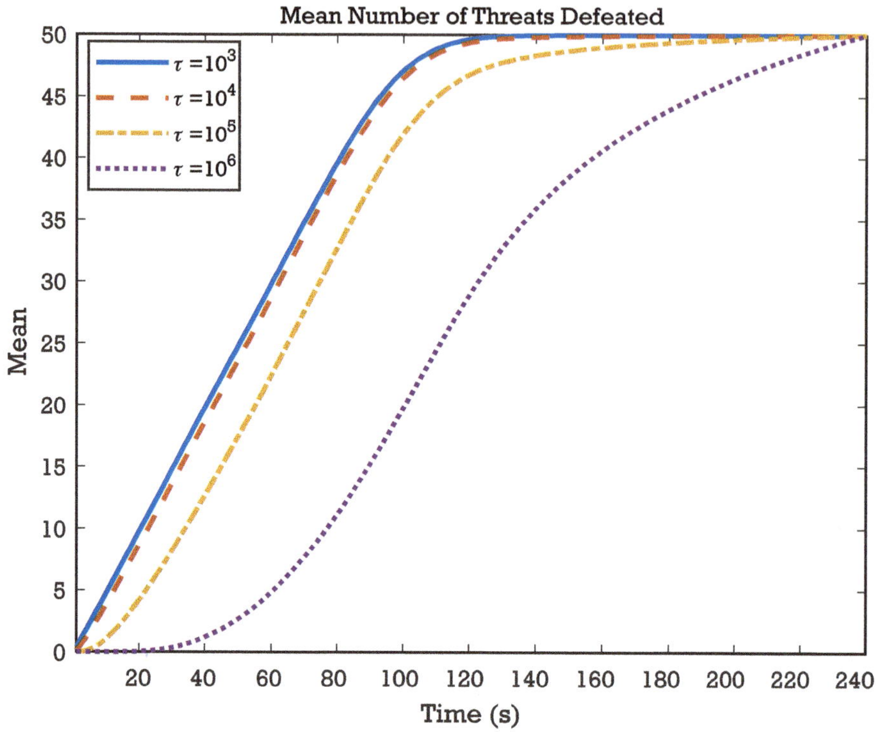

Figure 4.10 A plot under similar conditions as that in Figure 4.9, except the HPRF power level has been increased to 10^{12} W. This means that on average all the threats will be defeated within the given time period.

the HPRF DEW, in which case they may detonate an embedded explosive to destroy the DEW.

It is assumed that the threats are launched sequentially with a time delay. Hence the HPRF DEW will act to neutralize one threat at a time. Therefore this scenario can be modeled through a single server queue, as introduced in Chapter 2. The queueing model with reneging will only be examined in the following. With reference to the development in Section 2.2.2, it will be required to estimate the complementary distribution function (2.36), which will consequently involve estimation of the delays in the queue through the recursive expression (2.35).

Recall that $M(t)$ is the number of threats defeated by time t in the queueing system. Since threats may renege, this random variable also includes reneged threats, and so the point of view taken is that any threat that exits in the queue has been defeated, even if it expired due to waiting too long for service in the queue.

Suppose that the threats arrive through a renewal process with arrival times $\{S_n, n \in \mathbf{N}\}$, and that these threats travel toward the HPRF DEW in a straight trajectory such that the nth threat travels at a speed of v_n m/s, and would take T_n seconds to reach the target if not intercepted. Then $R_n(t) = \max\{0, v_n(T_n - t)\}$ is the distance of the target from the HPRF DEW at time t after its appearance in the combat space.

When the threat is in the queue it will experience a delay D_n before it reaches the server, unless it reneges. Under the assumption that it does not renege, the power density (3.8) must be integrated over the observational window $[S_n + D_n, S_n + D_n + t]$, which will then enable determination of the probability that the threat has been defeated by time t. Hence the latter is relative to its arrival at the server. Thus the probability that the nth threat is defeated by time t seconds is

$$\phi_n(t) = \mathbf{P}\left(\int_{S_n+D_n}^{S_n+D_n+t} I(s)ds > \tau\right), \tag{4.25}$$

where $I(s)$ is given by (3.8) and τ is the disruption threshold.

What is required for (2.36) is the random variable Y which is the service time of the target. If Y_n is the service time of the nth threat once it has entered service then as shown in [4] it follows that

$$\mathbf{P}(Y_n > t) = 1 - \phi_n(t), \tag{4.26}$$

and so (4.25) can now be applied to (4.26). In particular, if the resultant expression can be simplified it may be possible to produce a useful expression for simulation of Y_n through the inverse distribution function method.

In order to achieve the latter, suppose that each target is homogeneous such that A_R has an exponential distribution with parameter μ as utilized previously. Then by evaluating (4.25) one can show that the random variable Y_n may be expressed in the form

$$Y_n = \frac{T_n - S_n - D_n}{1 + \frac{\kappa \log(R_n)}{\mu\tau v_n^2(S_n+D_n-T_n)}}, \tag{4.27}$$

where R_n has a uniform distribution on the unit interval and $\kappa = \frac{P_D A_D}{\lambda^2}$.

In order to determine (2.36) it is required to specify a particular renewal process as well as a series of corresponding reneging times. Hence in the following a Poisson process will be assumed for the arrival process and exponentially distributed lifetimes of threats will be imposed. These assumptions, together with (4.27), then permit the Monte Carlo estimation of (2.36).

Suppose that each threat has $\nu_n = 40$ and $T_n = 120$, and that $\mu = 1$. The HPRF DEW operates at a wavelength of 1 GHz with an output power of 1 GW and antenna effective area of $A_T = 1\ \text{m}^2$. Assume that the threat vulnerability threshold is $\tau = 10^4\ \text{W/m}^2$. Hence in the numerical examples to follow, the mean threat arrival rate as well as their average lifetimes will be varied.

Figure 4.11 is for the case where both the mean arrival rate and threat reneging mean time are set to unity. The simulation was run only until 15 threats were generated, and this figure shows three complementary

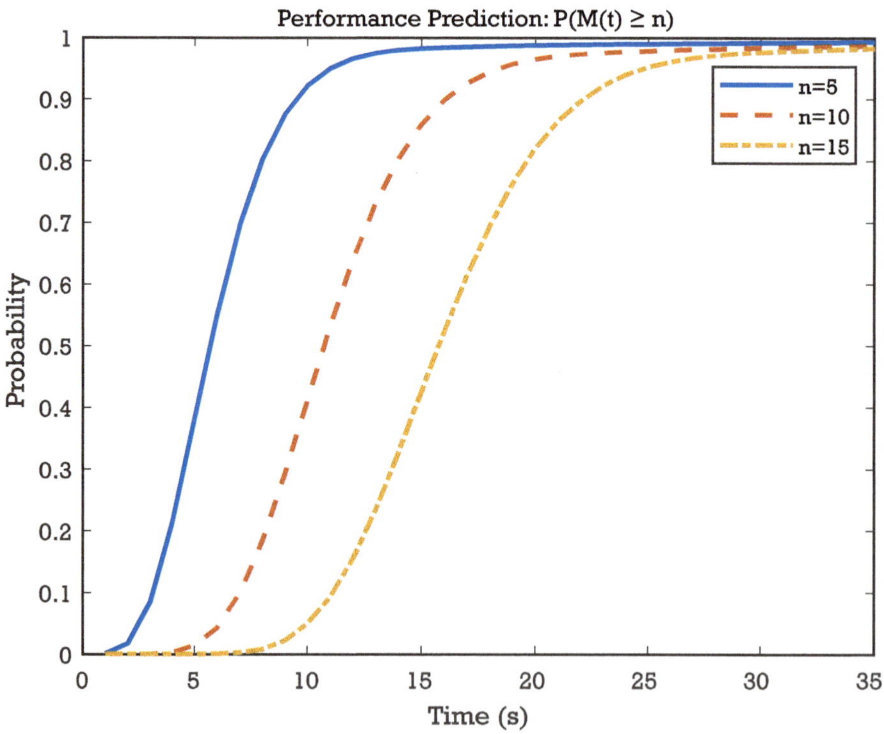

Figure 4.11 Case 1: Arrival rate and threat lifetime have mean unity.

distributions functions, namely $\mathbf{P}(M(t) \geq 5)$, $\mathbf{P}(M(t) \geq 10)$, and $\mathbf{P}(M(t) \geq 15)$. The Monte Carlo simulation consisted of 10^5 runs to generate each probability; this number of simulation runs will be employed throughout. The plot shows that in a very short time most of the threats are likely defeated.

The average reneging probability was also estimated in this case and it was found to be 0.9758. This implies that threats are more likely to renege in this situation, which also contributes to the apparent success of the HPRF DEW in defeating the threats rapidly. However, from the point of view of threat elimination, reneging is a successful outcome.

In order to understand the success of the HPRF DEW, the simulation in Figure 4.12 increases the threat lifetime mean to 50, yielding an average reneging probability of 0.6801, which is significantly smaller than the previous case. As the figure shows, it takes longer to ensure the threats are

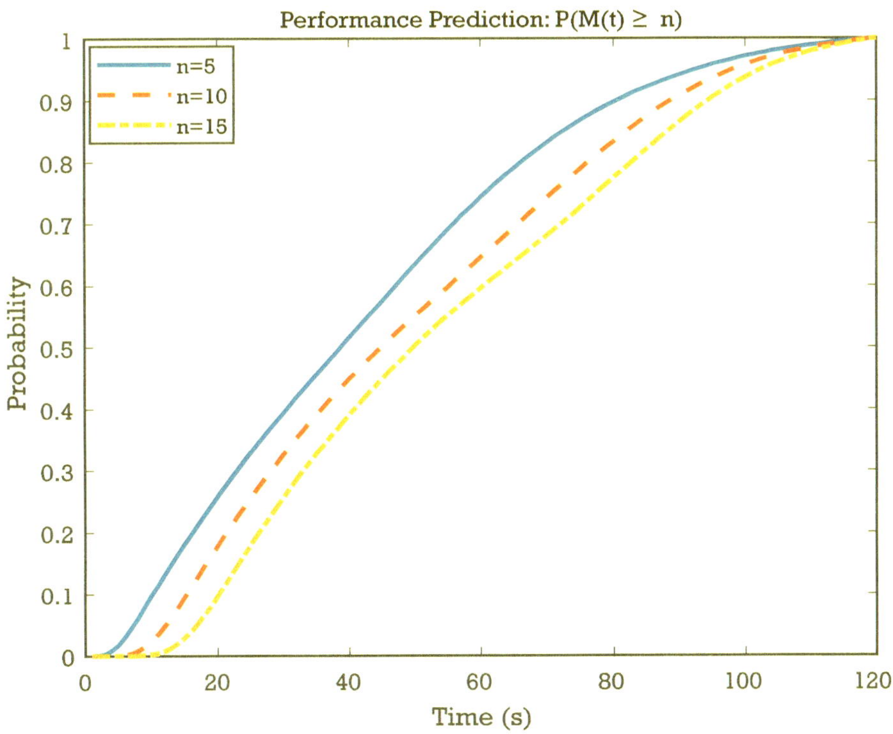

Figure 4.12 Case 2: Same arrival rate of unity but threats have a lifetime rate of 50.

defeated in the plotted time interval. Nonetheless the figure shows all the threats are likely eliminated by time 120s.

It is of interest to examine the impact of increasing the mean arrival rate, and so in Figure 4.13 this rate has been set to 10 while the threat lifetime rate has been reset to unity. Here the estimated average reneging probability is 0.4662. The figure is showing that while it is likely at least 5 threats are defeated by time 120s, the remaining threats will be taking longer to defeat, since they are remaining in the queue waiting to be served.

A useful way in which these three cases can be compared is to produce a plot of the mean number of threats eliminated over time, and so Figure 4.14 plots the three mean functions. Case 1 is for the mean arrival and threat lifetime rate set to unity; Case 2 corresponds to increasing the mean threat lifetime to 50. Case 3 is for the situation where the arrival rate is 10 and

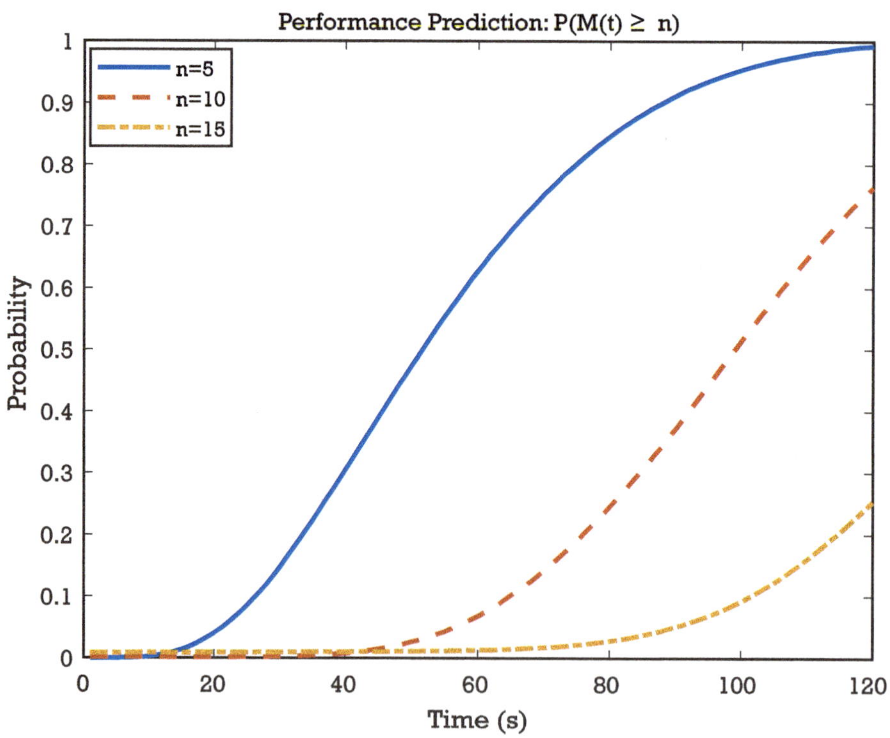

Figure 4.13 Case 3: Arrival rate of 10 with unity for lifetime rate.

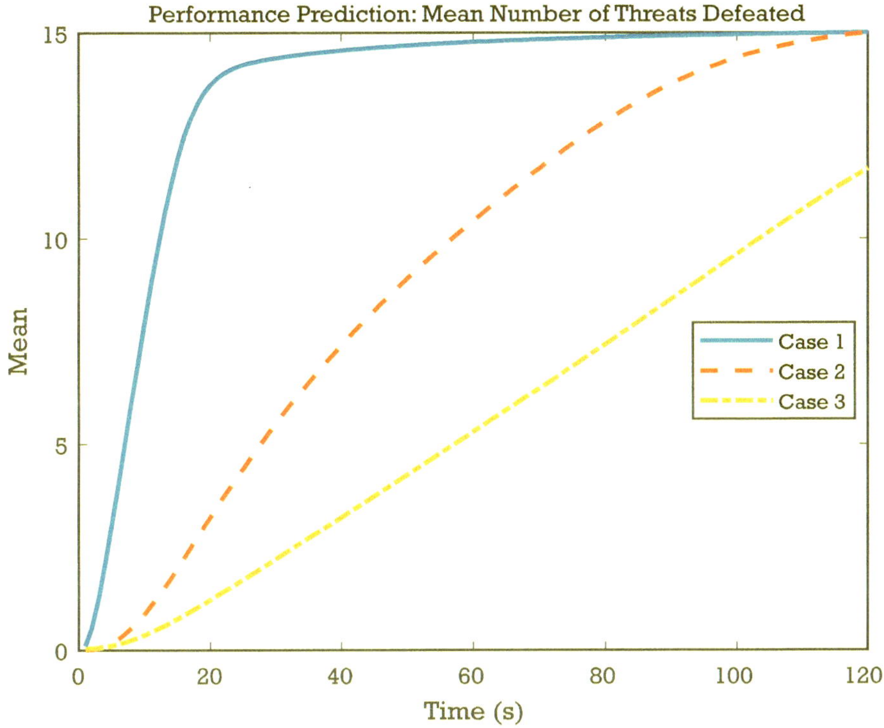

Figure 4.14 Plots of the mean of the number of threats defeated for the three cases considered. This figure facilitates a comparison between cases.

the threat lifetime is reset to unity. The plot validates the previous conclusions and provides a more useful way in which to compare the three cases illustrated.

4.3 Concluding Remarks

This chapter now completes the examination of HPRF DE system performance prediction. Using the power density functions derived in Chapter 3 a series of modeling situations were considered, ranging from one-to-one engagement of an HPRF DEW and a single threat, to the situation of a HPRF effector facing a multitude of threats. For the latter, the threats were assumed to either arrive simultaneously into the DEW's operational space or to arrive through a renewal process. In the final case, the threats were supposed to arrive in a linear trajectory toward the HPRF DEW but delayed in time. A series of relevant solutions were proposed to provide a means to gauge the performance of the HPRF DEW in defeating these threats under the arrival situations proposed.

Although in a majority of the cases a narrowband HPRF DEW effector was assumed, it was outlined how the developments from the preceding chapter could be used to produce a performance prediction capability for the case of a wideband or ultrawideband HPRF source. Chapter 7 will include a brief discussion of some of the future research directions that could be pursued to complement the analysis provided in this chapter.

References

[1] Weinberg, G. V., "Determining Performance Limits for Directed Energy Weapons in Collaborative Active Protection Systems," in *24th International Conference on Modelling and Simulation,* 2021, pp. 911–917.

[2] Weinberg, G. V., "Prediction of UAV Swarm Defeat with High-Power Radio Frequency Fields," *IEEE Transactions on Electromagnetic Compatibility,* Vol. 64, No. 6, 2022, pp. 2157–2162.

[3] Ross, S., *Stochastic Processes, 2nd Edition,* New York: John Wiley and Sons, 1996.

[4] Weinberg, G. V., "Performance Prediction of Directed Energy Weapons," *Progress in Electromagnetics Research M,* Vol. 108, 2022, pp. 79–88.

CHAPTER

5

Contents

5.1 Lasers: An Overview

5.2 Laser Damage Thresholds

5.3 Gaussian Beam Profiles

5.4 Irradiance Functions

5.5 Irradiance Function Examples

5.6 Some Final Comments on Irradiance Functions

References

The HEL Irradiance Function

This chapter begins the investigation of the performance prediction of HEL DEWs, and so it provides a brief introduction to the way in which laser performance can be assessed mathematically. The development will parallel that of the HPRF DEW case except the basis of performance prediction will be the laser irradiance function, which measures the laser beam power on a target per unit area. Hence it is exactly the power density function analyzed previously. Therefore constructing an appropriate irradiance function is the sine qua non of this chapter. As in the study of HPRF DEWs the underlying technology generating a HEL DE system beam is out of scope. However, environmental factors causing beam distortions, and consequently variations in the irradiance function, will be of fundamental importance, and examined in detail.

In a large number of cases a laser beam may be modeled through a Gaussian beam profile. The model's validity depends on the extent of diffraction, and as will become apparent, it is possible to include a laser beam quality factor into the model parameterization to extend the developments to non-Gaussian beam patterns in approximate terms. Under the assumption of a Gaussian beam it is possible to produce a closed-form expression for the irradiance distribution on a focal or target plane. The irradiance function can then be shown to decompose into two components. The first is a measurement of the peak intensity, which occurs at the beam's origin, while the second component distributes this peak power over a radial distance from the center of the beam but in a transverse plane. Much of the literature is focused on analysis of the peak irradiance, which can be used as a basis for the performance prediction of a HEL DEW. This underpins the assumption of a continuous HEL beam being applied onto the target until a required disruptive effect is achieved. A pulsed laser beam can also be used for target disruption purposes. A well-known example is provided by the fact that laser pointers have been used to cause temporary blindness in pilots. This soft-kill effect is referred to as laser dazzling [1]. Optical sensors may also be disrupted by such a low-power dazzling laser effect [2, 3]. These considerations will become important in terms of the problem of eliminating swarms of UAVs with HEL DEWs. This is an important point to consider because achieving a thermal effect on a small UAV may in fact be a redundancy. Hence not only will hard-kill thermal disruptions be of interest, but soft-kill options will also be of utility in the context of applications. Therefore, toward these objectives, a major requirement of this chapter is to produce useful expressions for the peak irradiance for application to the performance prediction of HELs in the next chapter. Before deriving these irradiance functions it is useful to provide a review of laser phenomenology as well as a discussion of laser-induced disruptive effects in targets.

5.1 Lasers: An Overview

As is well-known, a laser produces amplified light through the process of stimulated emission of radiation [4]. Stimulated emission of radiation is the process whereby a photon perturbs an excited electron resulting in it reducing to a lower energy state and consequently emitting a coherent photon of the same frequency, polarizations, and direction as the original perturbing electron [5]. In order to amplify light, electrons must be energized into this excited state that causes population inversion.

Due to the very small wavelengths (less than 10^{-6} m or a micron) a laser emits coherent light spatially allowing an intense beam of electromagnetic energy to be directed to a target in order to achieve an effect. Due to the collimated beam it is possible to cause an effect at considerable distances, making a HEL DEW an ideal long-range precision weapon [6].

Lasers are categorized via the medium through which the beam is produced [7]. Chemical lasers use chemical reactions while gas lasers utilize electrical current or optical pumping to excite gases. Solid-state lasers use solid-state materials as the gain medium, such as fiber lasers that use optical fibers. Although solid-state lasers are more stable and transportable than chemical and gas-based lasers, they tend to be lower in efficiency. There are also free-electron lasers that generate a stream of electrons that are passed through an alternating magnetic field to produce high energy beams, but are also subjected to a significant SWaP issue.

Lasers can operate in a pulsed or continuous mode, and under both modes can achieve this desired thermal disruption. A pulsed laser emits bursts of energy over discrete time intervals, which are known as the pulse duration. Continuous wave lasers apply energy continuously to a target. Short pulses with high peak power over long periods can have destructive effects, while long pulses will deliver damage to a target provided the peak power level is sufficient. Hence in either case if the peak power is sufficiently high the laser is likely to disrupt a target, provided it has sufficient time to cause a thermal effect. The time it takes to achieve a disruption on a target is known as the laser dwell time [8].

The difference in damage mechanisms between the two laser modes of operation is important to elucidate. As discussed in [9], in the case of pulsed lasers, the pulse duration is sufficiently short so that electric field effects are likely to cause system disruption. In the case of a continuous wave laser, since the laser is applied over a continuous period, target damage results from thermal effects. More recently, [10] demonstrated that in the continuous wave case, laser-induced damage is correlated with the optical surface's temperature at the center beam due to absorption.

For a HEL DEW to function successfully it requires high intensity beams to be propagated over considerable distances, while being subjected to a number of atmospheric attenuation factors. This beam must also be shaped using optical equipment, to ensure that it provides sufficient energy to the target. Adaptive optics have been proposed as a way in which to limit the effects of beam distortion through beam shaping based upon propagation considerations. A useful study of the integration of adaptive optics into the design of a HEL DEW is [5], which will be discussed subsequently after the development of irradiance functions.

The choice of wavelength for the HEL DEW is dependent on the environment in which it has been designed to operate. Hence the optimal choice of wavelength will be considerably different for land- and sea-based deployments. This is because of the fact that many of the laser beam distortion factors are wavelength dependent. Thermal blooming, for example, is an effect that results in defocusing or distortion of the laser beam [11]. This is caused by a small fraction of laser energy being absorbed by molecular and aerosol constituents of air, which causes heating in the air locally, resulting in a decrease in air density. As a consequence of this, the refractive index in the propagating medium, which measures the ability of light waves to bend within it, is modified [12]. Other significant beam distorting factors include turbulence, aerosol and molecular scattering, thermal scattering resulting from heated aerosols, and aerosol heating and vaporization [13]. It is important to note that aerosols can absorb laser energy thus resulting in significantly less peak power on a target. Useful references on the issues with laser beam propagation include [14] and [15].

There are many references useful from the point of view of this chapter's main theme. Some of these include [16] which examines fiber lasers for DEW applications, [17] which is concerned with the usefulness of HEL DEWs for short-range defense, [18] which investigates the integration of HELs into land-based mobile platforms, and [19] which considers the defeat of airborne threats through a HEL. The utility of HEL DEWs for counter UAV applications has been investigated in [20] and [21].

5.2 Laser Damage Thresholds

The purpose of this section is to provide a summary of some laser damage thresholds relative to vulnerabilities on targets. In particular, the targets considered will be UAVs, since these provide a useful suite of reference vulnerabilities from which one may extrapolate vulnerability levels on other types of threats.

A target's vulnerability will be measured by its disruption threshold, as in the case of HPRF DEWs. However, for laser effects the measurement of this threshold will depend on the mode of operation. For a pulsed laser the disruption threshold must be measured in terms of the energy in a pulse over the pulse duration. In the case of a continuous-wave laser the power density on the target per unit area is the appropriate measure of thermal effects. Since the focus will be on continuous-wave lasers, the disruption threshold will be measured in terms of irradiance.

UAVs can be equipped with IR cameras allowing them to perform ISR activities at night, and so there has been considerable interest in the application of lasers to disrupt IR cameras. As discussed in [21], the destruction of an optical detector can be achieved through a laser with a high energy pulse with short pulse length. For a continuous wave laser it has been reported that irradiance levels in the range of 10 to 100 kW per square cm for a 1s exposure can damage cameras. In [22] a quantum cascade laser operating at a wavelength of 10.6 μm and with a power level less than 100 mW is shown to induce damage on a camera with an irradiance of 5.9 W/m^2. In this case it was observed that 90% of the camera's pixels were saturated. When the irradiance was increased to 10 W/m^2, optical detector damage on the UAV was observed. These cases are for what is referred to as in-band damage, which means that the laser beam is directly focused into the camera's lens. Out-of-band damage is also possible to cause dazzling effects in sensors [21]. One important observation is that given the small irradiance thresholds required to disable a UAV's optical sensors, this soft-kill option may be preferable than trying to cause a thermal disruption, which also requires a more intense irradiation on the target. As a rough guide, effective dwell times are between 1 to 10 seconds, depending on the desired effect. Typical damage levels for melting result from irradiance levels of 0.01 to 10 kW per square cm (10^5 to 10^8 W/m^2), depending on the material. As a reference, damage limits for carbon fiber composite is 10^6 W/m^2. This is achievable with a laser operating with a power level of between 1 and 10 kW at a range of 1 km. Small UAVs can be disrupted with a lower level of power, in particular, between 1 and 5 kW.

A useful plot[1] in [21] shows a selection of material and irradiance levels to achieve thermal effects, as a function of laser pulse time. This is for a laser operating with a wavelength of 1.06 μm. Examples of results include steel requiring an irradiance of roughly 10^3 kW per square cm (10^{10} W/m^2) for a pulse of 10^{-6} s, which reduces to an irradiance of under 1 kW per square cm (10^7 W/m^2) for a 10s pulse. Aluminum requires almost an irradiance of 10^4 kW per square cm (10^{11} W/m^2) for a pulse length of 10^{-6} s to cause a disruption, which for a pulse length of 10s reduces to an irradiance of roughly

1. Figure 8 in [21].

1 kW per square cm. More common surface vulnerabilities on UAVs are carbon fiber reinforced composites (CFRC) and polystyrene materials, which are significantly more likely to be affected by heating. In particular, polystyrene will melt with an irradiance of approximately 50 kW per square cm with a pulse time of 10^{-6} s, while for a pulse time of 10s the required irradiance reduces to less than 10^{-2} kW per square cm (10^5 W/m^2). When CFRCs are subjected to a pulse of duration 10^{-6} s it will melt with an irradiance of just under 10^2 kW per square cm, while for a 10s pulse the required irradiance is just over 10^{-2} kW per square cm.

These results, extrapolated from [21], provide benchmark figures of vulnerability thresholds and average dwell times, which can be utilized in the performance prediction models introduced in the following chapter.

5.3 Gaussian Beam Profiles

As pointed out in Chapter 2, lasers operate in a TEM mode, and so diffraction causes light waves to spread transversely as they propagate. As such, the energy distribution of the laser beam is not uniformly spread in space, but is subjected to divergence. The laser irradiance function thus is designed to measure the laser power over a unit of area, which is perpendicular to the direction of propagation of the beam. The most fundamental form of laser beam is that which has a Gaussian beam profile, which is considered in this section.

The most commonly used helium-neon gas lasers operating in the fundamental TEM_{00} mode result in a Gaussian irradiance function [23]. Gaussian beam characteristics are typical of lasers that operate in single-mode, and while most modern lasers may have more complex beam patterns, the analysis of a Gaussian beam is of considerable importance. Such laser beam profiles are considered ideal, due to the fact that they correspond to a diffraction limited laser [24]. In the vicinity of the focal region of the beam, known as the waist, the beam is highly collimated with wavefronts planar and traveling in a transverse direction with zero beam divergence at its origin.

As discussed in [25], analysis of the propagation and focusing of Gaussian beams first appeared in [26], where it was demonstrated that two key laser parameters are sufficient for prediction of beam characteristics. These are the beam waist and its relative position in the direction of propagation [25]. The beam waist is the radius of the propagating beam that has a relative minimum value, usually at the focal point in the laser lens [27]. The latter reference also established a sufficient condition for a laser to exhibit a Gaussian beam profile. In particular, it was shown that the Gaussian irradiance distribution is maintained throughout a diffraction limited optical system provided that the beam is never truncated to less than twice the $\frac{1}{e^2}$ diameter [28].

The irradiance function of a Gaussian beam can be derived from the normalized modulus square of the electric field amplitude, where the latter arises as a solution to the paraxial Helmholtz equation [29]. This irradiance function is given by

$$I(r,z) = I_0 e^{-\frac{2r^2}{w(z)^2}}, \tag{5.1}$$

where r is the radial distance from center of the axis of propagation (the z-axis), z is the distance from the beam's focus, $w(z)$ is the radius of the beam's spot size at a distance z from the beam's focus, and $I_0 = I(0,0)$ is the intensity at the center of the beam's waist $(w(0) := w_0)$.

Figure 5.1 provides an illustration of the laser beam action. From an aperture in the corresponding aperture plane a collimated beam is applied to a spot in the focal plane, at a distance of $R(t)$, that has been indexed by time t.

Suppose that the beam has total power of P_0 at the focus or origin. For the beam centered on an aperture, the power passing through a circle of radius r in the transverse plain at a distance z from the focus is given by

$$P(r,z) = P_0 \left(1 - e^{-\frac{2r^2}{w(z)^2}}\right), \tag{5.2}$$

which can be obtained by integrating the irradiance function (5.1) over a circle of radius r. Based upon this it can be shown that

$$P_0 = \frac{1}{2}\pi I_0 w_0^2. \tag{5.3}$$

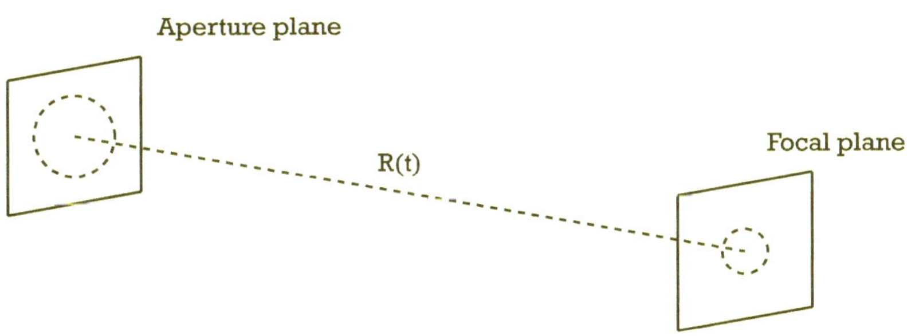

Figure 5.1 An illustration of the HEL propagation scene, where the laser beam originates from an aperture and is applied to a focal plane, at a distance $R(t)$ away.

Applying (5.3) to (5.1), one produces the irradiance function

$$I(r,z) = \frac{2P_0}{\pi w_0^2} e^{-\frac{2r^2}{w(z)^2}} \tag{5.4}$$

for a Gaussian beam profile.

Before examining the beam radius, it is necessary to discuss the Rayleigh range of the Gaussian beam. In contrast to conventional light beams, a Gaussian beam does not diverge linearly. Near the beam waist the beam divergence is small while in the far-field it is proportional to the wavelength.

The beam radius at a distance z along the transverse axis can be shown to be given by

$$w(z) = w_0\sqrt{1+\left(\frac{z}{z_R}\right)^2}, \tag{5.5}$$

where z_R is known as the Rayleigh range. Considering the solution to $w(z) = \sqrt{2}w_0$ one can conclude that the Rayleigh range is the point where the beam's radius has increased to $\sqrt{2}$ times its waist at the focus of the beam. An explicit expression for z_R is given by

$$z_R = \frac{\pi w_0^2 n_0}{\lambda}, \tag{5.6}$$

where n_0 is the index of refraction of the medium in which the beam is propagating.

As the beam propagates, the wavefront curvature increases to a maximum and then decreases to zero asymptotically. The Rayleigh range marks the boundary between the near-field and mid-field divergence. In addition to this it is the point where the wavefront curvature is maximum. The wavefront's curvature is given by the expression

$$C(z) = \frac{z}{z^2 + z_R^2}, \tag{5.7}$$

and the reciprocal of (5.7) is known as the radius of curvature. By an application of differential calculus it can be shown that (5.7) has a local maximum at $z = z_R$ as expected, and that the wavefront curvature increases from 0 at $z = 0$ to $\frac{1}{2z_R}$ at $z = z_R$. Beyond the Rayleigh range, the curvature decreases to zero. Calculations for laser spot sizes and other optical parameters will be inaccurate in the near or mid-fields. The far-field is for $z \gg z_R$ and parameter estimates must be taken in this region.

Figure 5.2 provides an illustration of the Gaussian beam, showing its waist w_0 at $z = 0$, the radius at a transverse distance of z, namely $w(z)$, the Rayleigh range, and the laser beam angle of divergence (5.8).

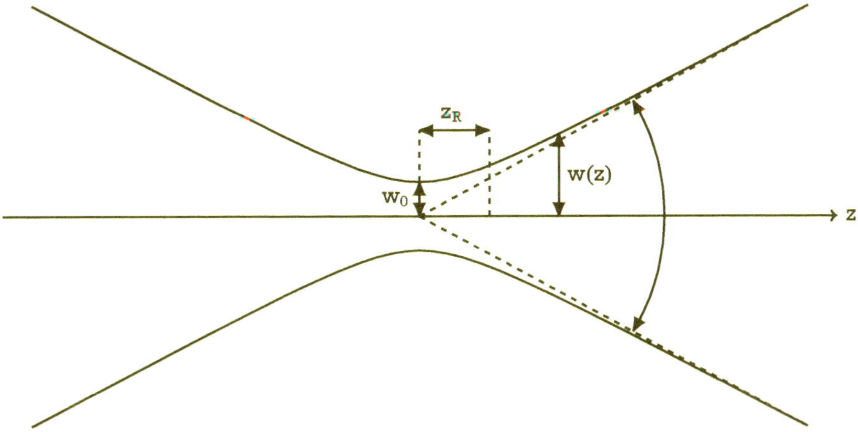

Figure 5.2 The Gaussian beam profile, where the wavefront begins at the focus at $z=0$ and propagates along the z-axis. Shown is the waist w_0 at $z=0$, the beam radius at a transverse distance of z, namely $w(z)$, the Rayleigh range, and the laser beam divergence angle.

As mentioned previously, a Gaussian beam experiences an angular divergence directly proportional to its wavelength. Based upon Figure 5.2 the angle of beam divergence is given by the solution to $\tan\left(\frac{\theta}{2}\right)=\frac{w(z)}{z}$, and so the angle is given by

$$\theta = 2\arctan\left(w_0\sqrt{\frac{1}{z^2}+\frac{1}{z_R^2}}\right). \tag{5.8}$$

One may take the limit of z (so in the far-field) as it becomes unbounded and utilizing the continuity of the inverse tangent function over the positive domain (5.8) yields

$$\theta = 2\arctan\left(\frac{\lambda}{\pi n_0 w_0}\right) \approx \frac{2\lambda}{\pi n_0 w_0}, \tag{5.9}$$

where (5.6) has been applied to produce the equality, while the approximation in (5.9) results from the fact that $\lambda \ll 1$ for laser wavelengths.

Although a Gaussian beam profile can be used as an exemplar, a real laser will experience greater beam divergence than that given by (5.9). Hence the well-known M^2 beam quality factor compares performance of a laser beam with that of a diffraction-limited Gaussian beam. As discussed in [30] the divergence of an actual laser beam is given by

$$\theta = M^2\frac{2\lambda}{\pi n_0 w_0}. \tag{5.10}$$

This completes the brief survey of laser beam characteristics of Gaussian beam profiles.

5.4 Irradiance Functions

Our focus is now directed towards deriving some useful expressions for the peak irradiance function to be applied in the next chapter. The following will be based closely upon the developments reported in a number of earlier investigations. In particular, both [31] and [32] have useful discussions and derivations of the peak irradiance function. Specifically, [31] investigates diffraction patterns produced by focused laser beams and demonstrates how the effects of jitter can be included if both the jitter statistics and the diffraction pattern are assumed Gaussian. Thermal blooming effects are considered for Gaussian beam patterns in [32], who derives a useful expression for the peak irradiance, which will also be introduced in the following discussion. Additionally, [14] provides a useful variation of the results in [32] which also accounts for thermal blooming in the peak irradiance function. Finally, there is a relevant irradiance expression introduced in [12] and [13] for the peak irradiance function based upon data from operational laser systems. These two references also provide useful parameterizations for the irradiance functions, which will also be adopted in the analysis in the next chapter.

The first irradiance function to be considered has been taken from [31], and is the peak intensity in the focal plane which also accounts for a number of atmospheric distorting factors. This is derived under the assumption that Gaussian functions can not only describe the beam pattern but also model the beam dispersion on the target due to random jittering. The irradiance function is given by

$$I_1(t) = \frac{P_T}{2\pi R(t)^2(\alpha_D^2 + \alpha_J^2)}, \tag{5.11}$$

where

- P_T is the total power of the laser;
- $R(t)$ is the distance to the target (indexed by time t);
- α_D is the angular standard deviation of the Gaussian diffraction pattern;
- α_J is the angular standard deviation due to the jittering of the beam motion.

It is also explained in [31] that the diffraction pattern parameter is given by

$$\alpha_D = M^2 \delta \frac{\lambda}{D}, \tag{5.12}$$

where M^2 is the beam quality factor initially introduced in (5.10), D is the diameter of the laser aperture, and δ is a constant. The inclusion of the M^2 factor allows the extension of the irradiance function to model approximately the irradiance of non-Gaussian beams. The constant δ is shown in [31] to be given by

$$\delta = \frac{2}{\pi} \approx 0.450. \tag{5.13}$$

Beam jitter can be measured through the mean square radial displacement of the focal spot, so that in (5.11) α_J^2 is the variance of the single-axis jitter angle. This basic irradiance model can be then extended, as reported in [32], to include atmospheric transmittance and turbulence factors. In particular the peak irradiance function is given by

$$I_2(t) = \frac{P_T \mathcal{T}(t)}{2\pi R(t)^2 \left(\alpha_D^2 + \alpha_T^2 + \alpha_J^2\right)}, \tag{5.14}$$

where $\mathcal{T}(t)$ is the atmospheric transmittance and α_T is the angular standard deviation of the Gaussian diffraction pattern, accounting for the effects of atmospheric turbulence on the beam, and all other parameters are as defined in the specification of (5.11).

The atmospheric transmittance function introduced above is given by

$$\mathcal{T}(t) = e^{-\kappa_T R(t)}, \tag{5.15}$$

where κ_T is the extinction coefficient, accounting for the total loss due to both absorption and scattering effects. This coefficient is usually decomposed into two components, so that $\kappa_T = \kappa_1 + \kappa_2$, where κ_1 corresponds to loss due to absorption while κ_2 accounts for loss due to scattering.

As pointed out in [14] these quantities have well-known tabulated results, based upon a number of experimental campaigns, and so may be considered constant under conditions of minimal beam turbulence. The term α_T in (5.14) is also referred to as the turbulence-induced beam spreading factor and approximations for it have been derived in [33]. In particular,

$$\alpha_T = \frac{\lambda}{r_0}, \tag{5.16}$$

where r_0 is known as the transverse coherence length, given by

$$r_0 = 0.4983\left[\frac{\lambda^2}{c_N^2 R(t)}\right]^{\frac{3}{5}}, \tag{5.17}$$

and c_N^2 is referred to as the refractive index structure constant [34], which quantifies laser beam atmospheric turbulence. As has been reported in a number of studies, weak turbulence corresponds to an index of the order 10^{-17} m$^{-2/3}$ while strong turbulence results in it of the order 10^{-13} m$^{-2/3}$ [5, 35]. An implicit assumption here is that this index remains constant over the beam path; variations on this assumption are provided in [19].

As discussed in [13], the effective strength of turbulence also depends upon the laser wavelength and the range of propagation. Hence the Rytov variance given by

$$\sigma_R^2 = 1.23 c_N^2 (2\pi\lambda)^{-7/6} R(t)^{11/6} \tag{5.18}$$

integrates these features into a a single measure of turbulence [36]. This measures the irradiance function scintillation level on axis at the target, and strong turbulence occurs when $\sigma_R^2 > 1$ while weak turbulence corresponds to $\sigma_R^2 < 1$.

It is now important to introduce thermal blooming into the irradiance models, since it can have a significant impact on laser beam coherence on a target. To account for thermal blooming the approach in [32] is to multiply the peak irradiance by a factor that is the ratio of bloomed to unbloomed peak irradiance. This introduces a distortion number accounting for thermal blooming effects. An appropriate irradiance function, which accounts for thermal blooming, is given by

$$I_3(t) = \frac{P_T \mathcal{T}(t)}{2\pi R(t)^2 \left(\alpha_D^2 + \alpha_T^2 + \alpha_J^2\right)\left(1 + 0.0625 N_B^2(t)\right)}, \tag{5.19}$$

where the term $1 + 0.0625 N_B^2(t)$ is a measure of thermal blooming and $N_B(t)$ is known as the distortion parameter. This term introduces a nonlinear complexity into the irradiance function, which will provide some challenges in the construction of performance prediction models accounting for thermal blooming. Under the assumption of a focused beam with no slewing, [32] provides an approximation for the distortion parameter, while [5] has a clearer expression for it which is adopted in the following. The approximate distortion parameter is given by

$$N_B(t) \approx \frac{8\sqrt{2}\pi(n_T)\kappa_1 P_T \mathcal{T}(t)R(t)}{n_0 \rho c_p \nu D\lambda}, \tag{5.20}$$

where n_0 is the atmospheric refractivity index, $n_T = \frac{dn}{dT}$ is the rate at which the index of refraction changes with temperature, κ_1 is the atmospheric absorption parameter, $\mathcal{T}(t)$ is the atmospheric transmittance, ρ is the air density, c_p is the air heat capacity, and ν is the cross-wind speed, relative to the axis of propagation. The other parameters in (5.20) have been defined previously. Hence the distortion parameter is dependent on the laser power linearly, demonstrating the fact that by increasing the laser power one cannot eliminate the effects of thermal blooming. The critical value of N_B is 25, since for distortion parameters beyond this limit the effects of thermal blooming become significant [5].

As an example of an irradiance function constructed from practical considerations, [13] provides a model constructed for the maritime domain. It is observed that laser beam propagation through the atmosphere results in spreading of the beam radius and wandering of the laser beam centroid, which results in variations in the laser irradiance on a target. Hence [13] incorporates variations of the laser beam radius on a target into their model development. Their peak irradiance function is given by

$$I_4(t) = \left(\frac{6.28 P_T}{M^4 + 2.9\left(\frac{R_o}{r_o}\right)^2 + (\pi R_o \alpha_J/\lambda)^2} \right) \left(\frac{R_o}{\lambda R(t)} \right)^2 e^{-\kappa_T R(t)}, \tag{5.21}$$

where R_0 is the initial laser spot size on the target, and the other parameters are as defined previously. Note that for a Gaussian beam pattern R_0 can be expressed in terms of the laser spot size $w(z)$ at a transverse distance of z from the beam's origin.

In addition to this, [12] provides useful model parameterizations across a number of different operational environments, namely maritime, desert, rural, and urban operational settings. These model parameterizations have been constructed in an attempt to determine optimal wavelength and power settings, across each environment, to maximize the efficiency of laser beam propagation.

5.5 Irradiance Function Examples

In order to provide some examples of the irradiance functions defined above it is necessary to stipulate appropriate model parameterizations. Hence for brevity only one operational environment will be considered, which will be maritime, since one can extrapolate model characteristics from [12]. In addition to this, some atmospheric parameters have been based upon results documented in [37]. Table 5.1 provides an enumeration, partially influenced from [12], of model characteristics for a maritime setting. In several cases three model parameters are included, which should be matched with other corresponding parameters with three available choices. Hence, for example, for a wavelength of 1.625 μm, one would utilize $\kappa_2 = 7 \times 10^{-2}$. The atmospheric refractivity index and change rate has been based upon Table II in [37], for a standard atmosphere at an altitude of 1 km above the Earth's surface, and by deriving a rate change for temperature in terms of altitude, and applying this to the estimated rate change of refractive index with altitude.

Hence Table 5.1 provides three sets of parameterizations for the irradiance functions. It is important to observe that the four irradiance functions (5.11), (5.14), (5.19), and (5.21) are deterministic functions, in contrast to those developed for the HPRF DEW cases examined in Chapter 3. Consequently one can, in principle, use these irradiance functions to determine the likely irradiance level on a target at a given distance for an appropriate set of laser and propagation parameters. However, it seems plausible to model the disruption threshold of a target using a random process, as also demonstrated in Chapter 3. This is a suitable approach because from the HEL DEW's

Table 5.1 Model Parameterizations, Influenced by [12], for a Maritime Operational Environment

Parameter and Units	*Values*
Laser wavelength λ (μm)	1.045, 1.625, 2.141
Laser spot size R_o (cm)	50
Aperture diameter D (cm)	80
Jitter α_J (radians)	$\ll 10^{-2}$
Laser beam quality M^2	1
Wind velocity (m/s)	5
Turbulence strength c_N^2 (m$^{-2/3}$)	10^{-15}
Aerosol scattering coefficient κ_2 (km^{-1})	1.2×10^{-1}, 7×10^{-2}, 5×10^{-2}
Aerosol absorption coefficient κ_1 (km^{-1})	2×10^{-3}, 2×10^{-3}, 3×10^{-3}
Atmospheric refractivity index n_0	1.0000286
Atmospheric refractivity change rate n_T (degree K^{-1})	6.77×10^{-6}
Air density ρ (kg/m^3)	1.2
Air heat capacity c_p (joules per kg per degree K)	700

perspective, it will not have a priori knowledge of a target's vulnerability threshold, but there will be likely estimates of threshold levels which can then be used as a basis for the parameterization of a statistical model for the vulnerability threshold. Therefore, this idea will be adopted and explored in the next chapter. Numerical examples of these irradiance functions will now be investigated relative to the model parameterizations in Table 5.1.

Consider (5.11) for the scenario where $P_T = 1$ kW, $D = 0.8$m, $M^2 = 1$, and suppose that there is no beam jitter $\alpha_J = 0$. Figure 5.3 shows the irradiance function with the three wavelengths selected from Table 5.1. As can be observed, for this case there is irradiance of the order of 10^7 W/m^2 at a distance of 5 km for all three wavelengths. Hence under this idealized situation the laser will be an effective long-range strike weapon against a range of targets. To demonstrate the impact jitter can have on the irradiance function, Figure 5.4 replicates the plot of Figure 5.3 with a jitter coefficient of

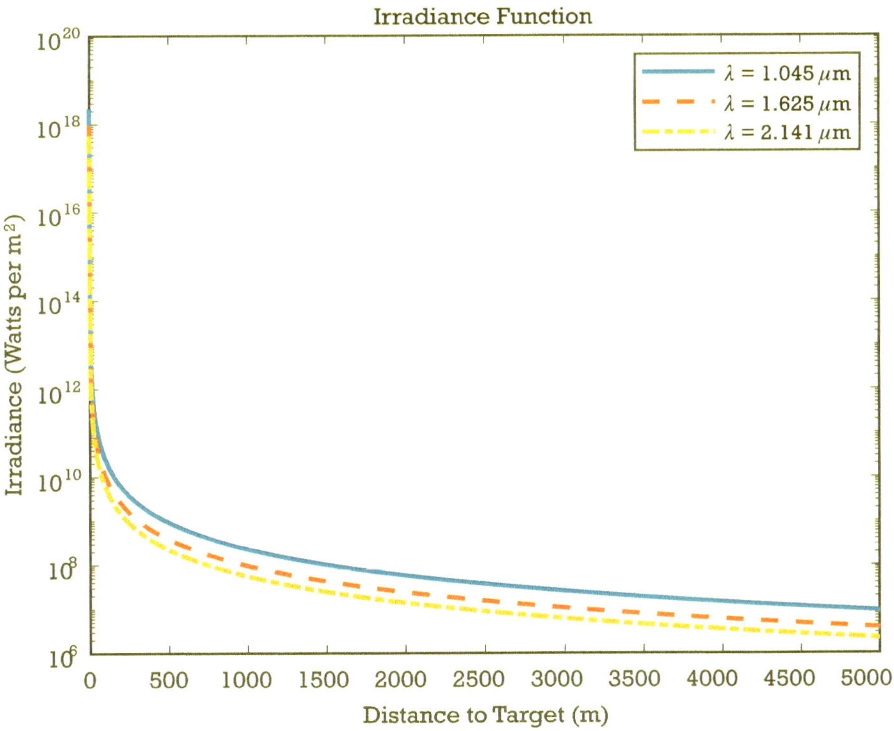

Figure 5.3 A plot of the irradiance function (5.11) with no beam jitter and a power level of 1 kW.

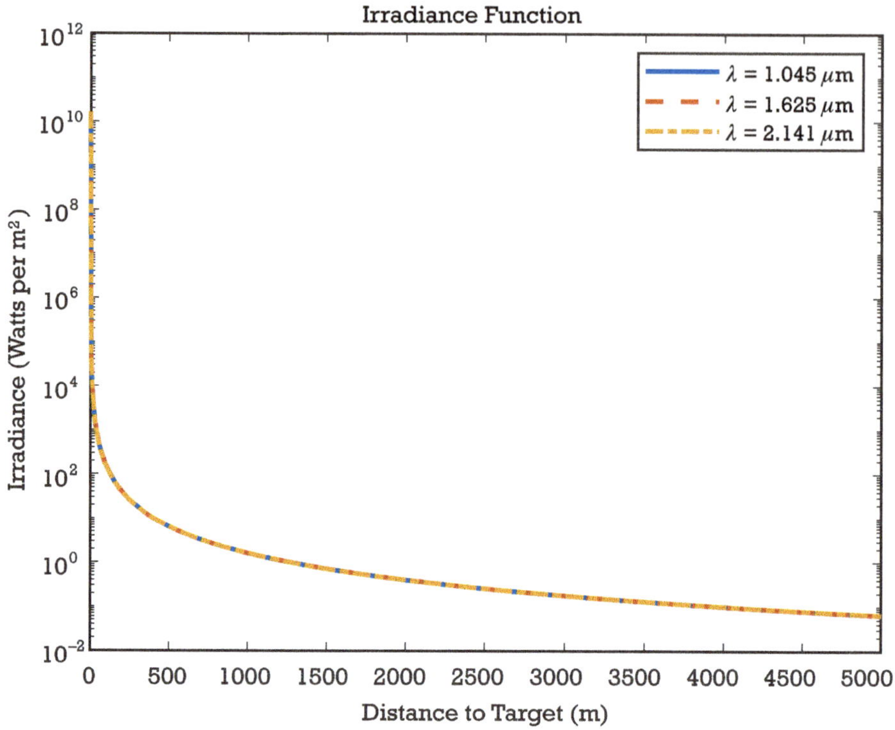

Figure 5.4 The irradiance function (5.11), under the conditions used to generate Figure 5.3, except that jitter of the order 10^{-2} radians is introduced.

$\alpha_J = 10^{-2}$ radians. As can be observed, there is a significant reduction of the irradiance with distance. Note that this is only a small level of jitter. When it is increased beyond 1 radian there is extreme degradation in irradiance. Figure 5.4 also shows that there is no significant difference in performance over each wavelength.

The problems with jitter can be remedied at shorter distances if the laser power can be increased. Figure 5.5 shows the situation with the same level of jitter but an increase in power to $P_T = 100$ kW. Here there is an improvement in irradiance, although no variation in the three wavelength cases. While this remedies the effects of jitter in this example, it is not always practical to increase the laser power to improve overall performance, as will be demonstrated with some examples to follow in the next section.

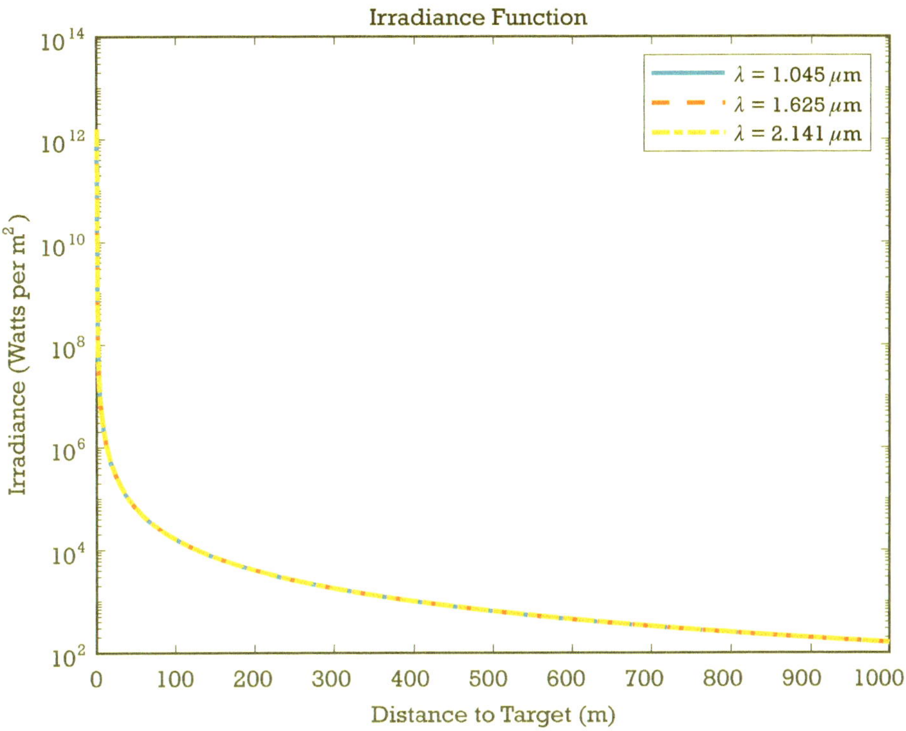

Figure 5.5 Improving the irradiance by increasing the laser power to 10^5 W, while retaining the jitter and other parameters employed in Figures 5.3 and 5.4.

Before examining the irradiance function (5.14) the Rytov variance is illustrated for the two extremities where $c_N^2 = 10^{-13}$ (strong turbulence) and $c_N^2 = 10^{-17}$ (weak turbulence). Figure 5.6 is for the strong turbulence case, and as the figure shows, it is clear that the effects of turbulence will be significant at longer ranges. For up to roughly 500m the Rytov variance is less than unity, and the largest Rytov variance values occur for a wavelength of $\lambda = 1.045$ μm. Figure 5.7 demonstrates that in the weaker turbulence case the Rytov variance is significantly smaller than unity even at a range of 5 km.

Next the irradiance (5.14) is considered, with $P_T = 1$ kW, $D = 0.8$m, $M^2 = 1$, and $c_N^2 = 10^{-15}$. The latter value for the turbulence coefficient corresponds to semistrong turbulence. Figure 5.8 is for the case where jitter is

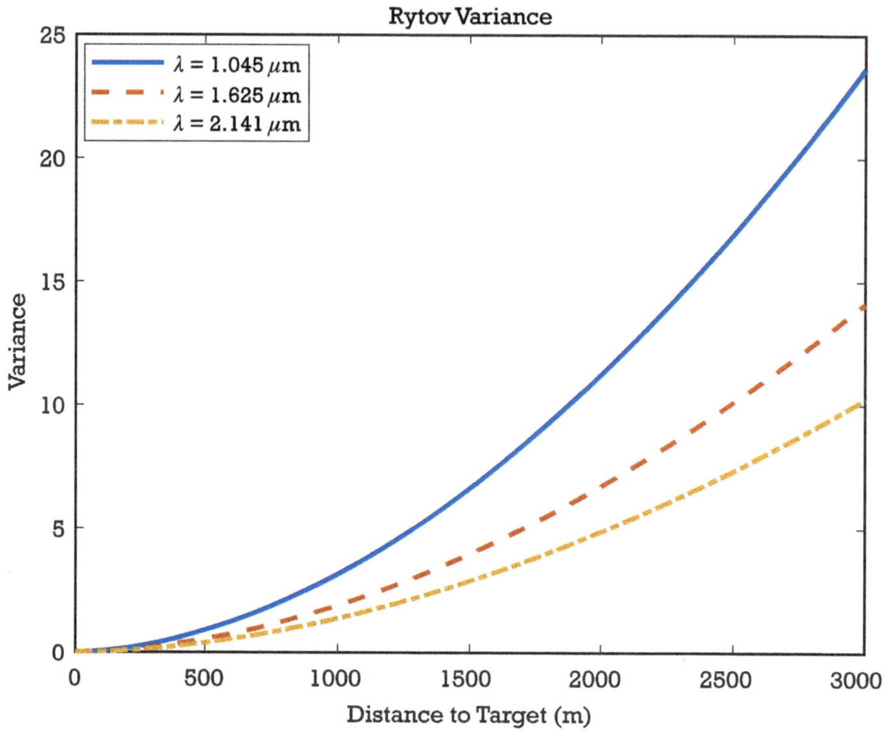

Figure 5.6 The Rytov variance in the case of strong turbulence with $c_N^2 = 10^{-13}$.

not present. In this example there is still considerable irradiance at a distance of 5 km (of the order of 10^5 W/m^2). Over the three wavelength cases there is a slight somewhat fluctuating deviation in performance. When jitter is included there is a similar decline in irradiance as previously demonstrated. This can also be improved at shorter ranges by increasing the laser power. Figure 5.9 demonstrates this by increasing the laser power to $P_T = 100$ kW when $\alpha_J = 10^{-2}$. At a distance of 1 km the irradiance is of the order of 100 W/m^2 in this case for all three wavelengths.

Examples including thermal blooming are now investigated, and so (5.19) is considered, together with the distortion number (5.20). For this situation the parameterization in Table 5.1 is adopted unless otherwise stated. In the first instance suppose that $P_T = 1$ kW and there is no jitter. Here one observes that the irradiance decreases rapidly with distance, but at shorter

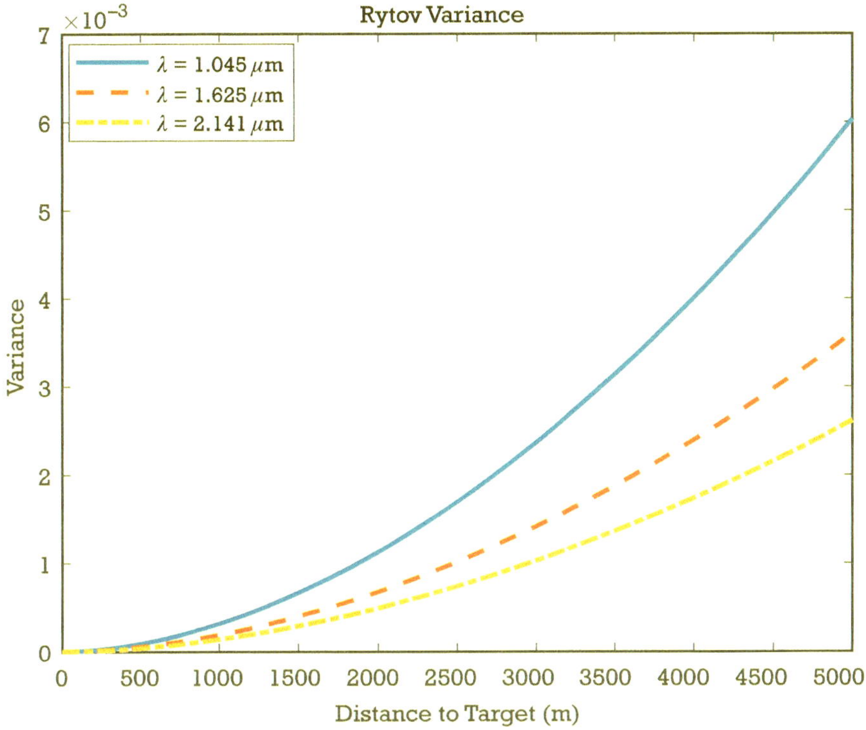

Figure 5.7 Weak turbulence and the Rytov variance for the case where $c_N^2 = 10^{-17}$.

ranges it will be considerate, as illustrated in Figure 5.10. For example, at a distance of 150m the irradiance is approximately 5×10^4 W/m^2. To gain an insight into the impact of thermal blooming in this scenario, Figure 5.11 plots (5.20), which shows that the effects of thermal blooming are significant. It is informative to examine the same example with an increase in laser power. Hence Figure 5.12 replicates the example except where the power has been increased to $P_T = 1$ MW. Here there is a much more significant decrease in the irradiance; this is also in the absence of jitter. Figure 5.13 demonstrates that the increase in laser power has caused the thermal blooming effects to increase, which has resulted in the decrease in irradiance. Therefore it is clear that increasing the laser power will not remediate the effects of thermal blooming, but can instead cause a degradation in irradiance.

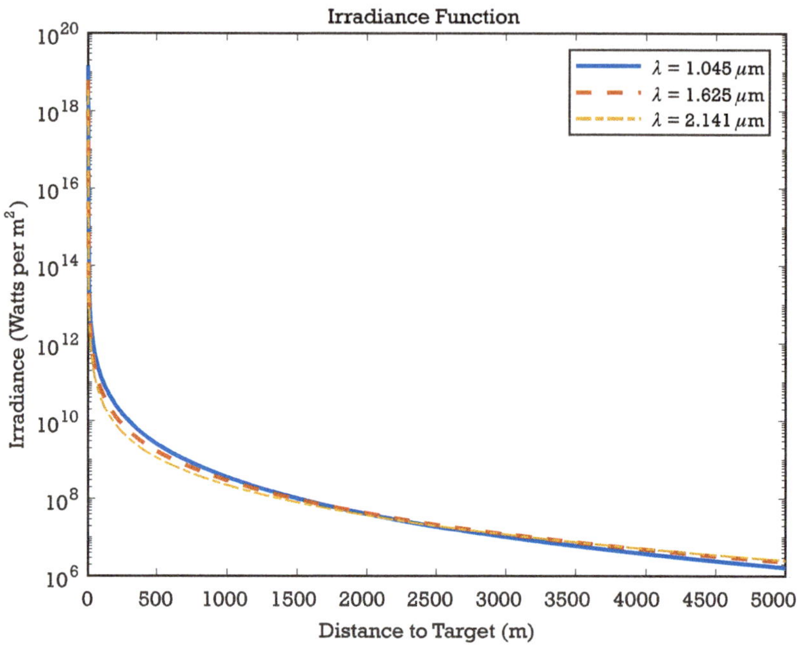

Figure 5.8 Irradiance function (5.14) in the presence of turbulence without jitter.

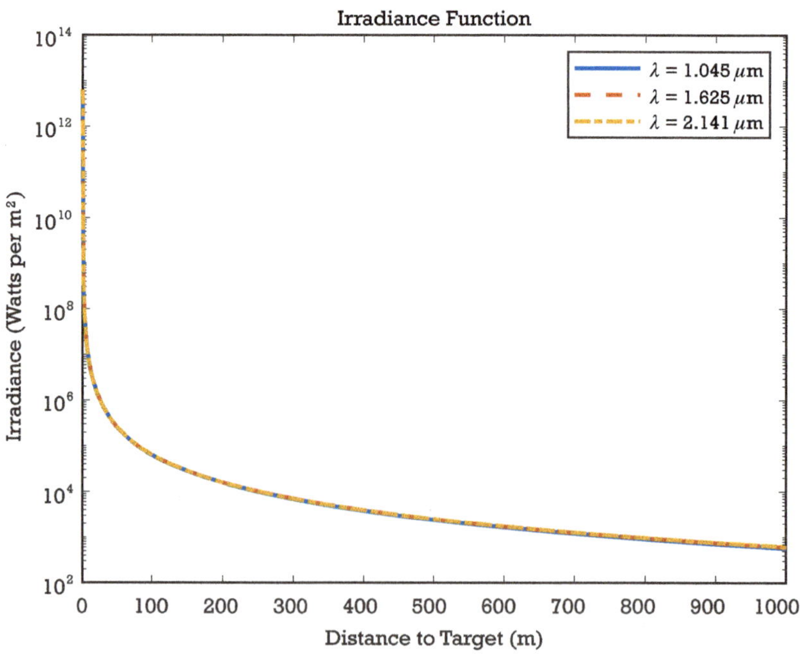

Figure 5.9 Increasing the laser power to compensate for jitter.

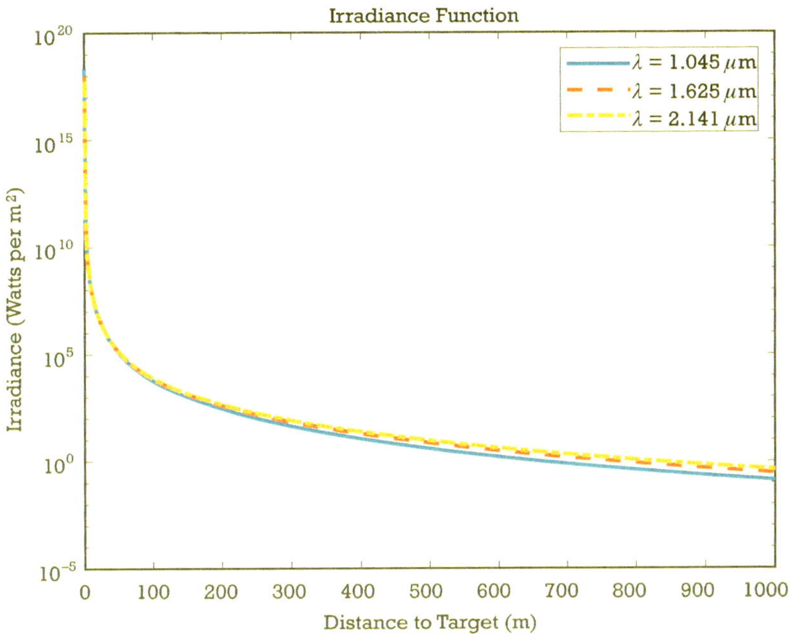

Figure 5.10 Example of the irradiance function (5.19).

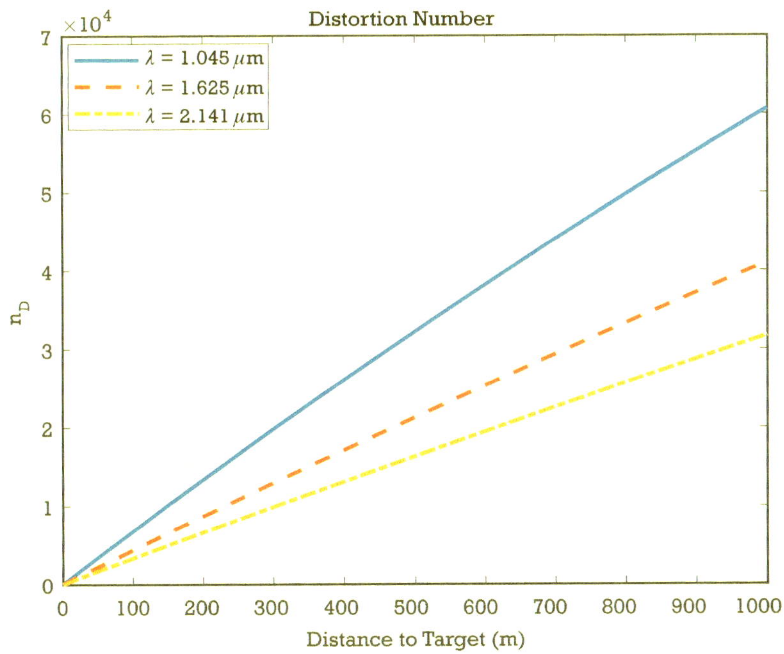

Figure 5.11 Distortion number (5.20) for Figure 5.10.

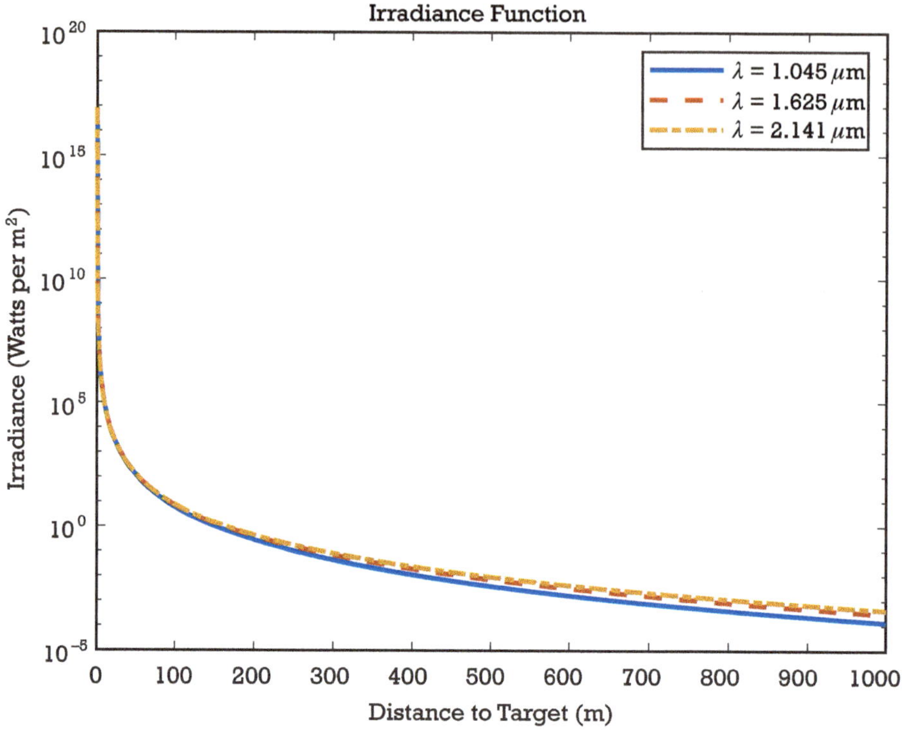

Figure 5.12 The irradiance function (5.19).

Finally, the irradiance function (5.21) is examined, again with model parameterizations provided by Table 5.1 unless otherwise stated. Figure 5.14 is for the case of $P_T = 1$ kW with no jitter. The figure shows slight fluctuations between the results for different wavelengths and that strong irradiance is present at 5 km. When jitter is included this results in a significant decline in the irradiance as in previous examples.

5.6 Some Final Comments on Irradiance Functions

The purpose of this chapter was to introduce a series of potential irradiance functions to be used in the following chapter, which will explore performance prediction of HEL DEWs. Hence the models (5.11), (5.14), (5.19), and (5.21) were introduced, and some examples were provided to illustrate their performance. A notable result was that (5.19) demonstrated that increasing

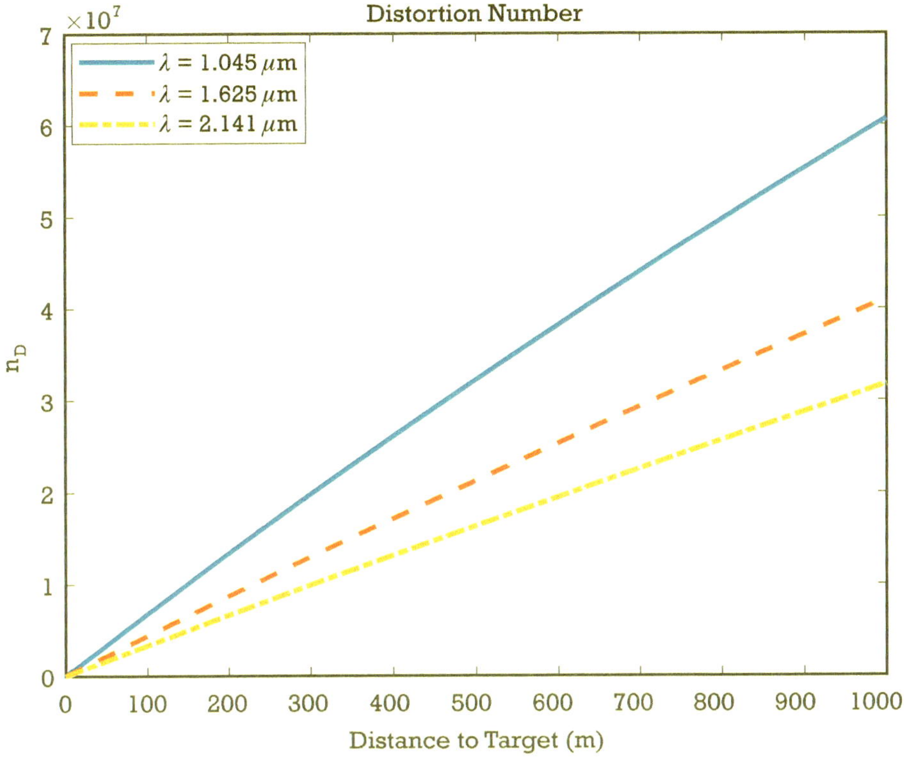

Figure 5.13 Distortion number (5.20) corresponding to Figure 5.12.

laser power will not remedy the effects of thermal blooming, and so as an approximate rule of thumb, lasers operating up to the kW power level will only be examined in the following. As remarked earlier in this chapter, adaptive optics has been proposed as a suitable way in which to address the problem of turbulence. Due to the fact that this involves wavefront detection considerations and laser system design, a brief description of it is deferred to Appendix B.

A review of the methodology in Chapter 4 will indicate that it will be of interest to integrate the appropriate irradiance functions, over an observational window, in order to quantify laser performance. Hence it is clear that if the more useful irradiance functions are utilized (namely (5.19) or (5.21)) then the process of numerical integration will be time-consuming. Hence in

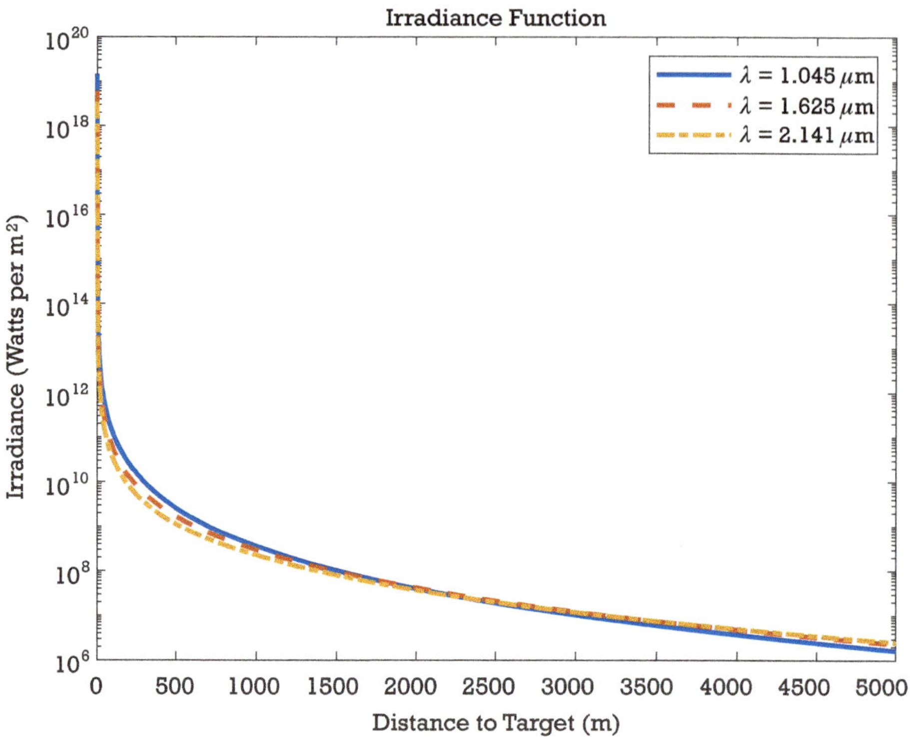

Figure 5.14 The irradiance function (5.21) with $P_T = 1$ kW and no jitter.

the following chapter it will be necessary to express any irradiance function applied in the simplest form to facilitate the numerical analysis.

This may also be achieved by limiting the scope of the applications in the following chapter. Hence if one restricts attention to the maritime domain, where a ship-based HEL DEW is defending against a series of UAVs, then one can apply the parameterization in Table 5.1 to an irradiance function and examine whether some components can be eliminated. If it is supposed that the UAVs have a significantly small detectability (small size or low visibility from a radar perspective) one may suppose that engagements take place within a 1 km radius. If the HEL DEW is operating in such short ranges then the effects of atmospheric transference may be considered negligible, thus eliminating this characteristic from the corresponding irradiance functions. These assumptions will be explained and justified in Chapter 6.

References

[1] Hueber, N., et al., "Analysis and Quantification of Laser-Dazzling Effects on IR Focal Plane Arrays," in *SPIE Proceedings*, Vol. 7660, 2010.

[2] Durecu, A., P. Bourdon, and O. Vasseur, "Laser-Dazzling Effects on TV Cameras: Analysis of Dazzling Effects and Experimental Parameters Weight Assessment," in *SPIE Proceedings*, Vol. 6738, 2007.

[3] Ritt, G., "Laser Safety Calculations for Imaging Sensors," *Sensors*, Vol. 19, 2019, p. 3765.

[4] Milonni, P. W., and J. H. Eberly, *Laser Physics*, Hoboken, NJ: John Wiley & Sons, 2010.

[5] Puent, D., "Integration of Adaptive Optics into High Energy Laser Modelling and Simulation," Technical report, Naval Postgraduate School, Monterey, California, 2017.

[6] Nielsen, P. E., *Effects of Directed Energy Weapons*, Washington, D.C.: National Defense University, 1994.

[7] Zhang, Y. X., et al., "Damage to Aircraft Composite Structure caused by Directed Energy System: A Literature Review," *Defence Technology*, Vol. 17, 2021, pp. 1269–1288.

[8] Burley, J. L., "Comparison of High Energy Laser Expected Dwell Times and Probability of Kill for Mission Planning Scenarios in Actual and Standard Atmospheres," Technical report AFIT/APPLYPHY/ENP/12M02, Air Force Institute of Technology, 2012.

[9] Palmer, J. R., "Continuous Wave Laser Damage on Optical Components," *Optical Engineering*, Vol. 22, No. 4, 1983, pp. 435–446.

[10] Slinker, K., et al., "Determining and Scaling Continuous-Wave Laser-Induced Damage Thresholds of Thin Reflectors," *Optics Express*, Vol. 27, No. 4, 2019, p. 4748.

[11] Penano, J., P. Sprangle, and B. Hafizi, "Propagation of High-Energy Laser Beams Through Atmospheric Stagnation Zones," *Journal of Directed Energy*, Vol. 2, 2006, pp. 107–117.

[12] Sprangle, P., J. Penano, and B. Hafizi, "Optimum Wavelength and Power for Efficient Laser Propagation in Various Atmospheric Environments," Technical report, Naval Research Laboratory, Washington, D.C., 2005.

[13] Sprangle, P., J. Penano, and B. Hafizi, "Beam Combining and Atmospheric Propagation of High Power Lasers," Technical report, Naval Research Laboratory, Washington, D.C., 2011.

[14] Gebhardt, F. G., "High Power Laser Propagation," *Applied Optics*, Vol. 15, No. 6, 1976, pp. 1479–1493.

[15] Cook, J. R., "Atmospheric Propagation of High Energy Lasers and Applications," in *AIP Conference Proceedings*, American Institute of Physics, Vol. 766, 2005, pp. 58–72.

[16] Sprangle, P., et al., "High-Power Fiber Lasers for Directed-Energy Applications," Technical report, Naval Research Laboratory, Washington, D.C., 2008.

[17] Lavan, M., "High Energy Laser Systems for Short Range Defense," *Acta Physica Polonica-Series A General Physics*, Vol. 115, No. 6, 2009, p. 959.

[18] Hafften, M., and R. Stratton, "High-Energy Laser Weapon Integration with Ground Vehicles," Technical report, United Defense LP Minneapolis MN Advanced Programs, 2004.

[19] Yun, Q., B. Song, and Y. Pei, "Modeling the Impact of High Energy Laser Weapon on the Mission Effectiveness of Unmanned Combat Aerial Vehicles," *IEEE Access*, Vol. 8, 2020, pp. 32246–32257.

[20] Steinvall, O., "The Potential Role of Laser in Combating UAVs: Part 1: Detection, Tracking and Recognition of UAVs," in *Proceedings of SPIE*, 2021.

[21] Steinvall, O., "The Potential Role of Laser in Combating UAVs: Part 2: Lasers as a Countermeasure and Weapon," in *Proceedings of SPIE*, 2021.

[22] Lewis, G. D., et al., "In-Band Low-Power Laser Dazzle and Pixel Damage of an Uncooled LWIR Thermal Imager," in *SPIE Proceedings*, Vol. 10797, 2018.

[23] Kogelnik, H. W., and T. Li, "Laser Beams and Resonators," *Applied Optics*, Vol. 5, 1966, p. 1550.

[24] Shimoda, K., *Introduction to Laser Physics*, Springer Series in Optical Sciences, Boston, MA: Springer, 1984.

[25] Herman, R. M., J. Pardo, and T. A. Wiggins, "Diffraction and Focusing of Gaussian Beams," *Applied Optics*, Vol. 24, No. 9, 1985, pp. 1346–1354.

[26] Fox, A. G., and T. Li, "Resonant Modes in a Maser Interferometer," Technical Report, Bell Systems Technology, Vol. 40, 1961, p. 453.

[27] Dickson, L. D., "Characteristics of a Propagating Gaussian Beam," *Applied Optics*, Vol. 9, No. 8, 1970, pp. 1854–1860.

[28] Hirleman, E. D., and W. H. Stevenson, "Intensity Distribution Properties of a Gaussian Laser Beam Focus," *Applied Optics*, Vol. 17, No. 21, 1978, pp. 3496–3499.

[29] Svelto, O., *Principles of Lasers*, New York: Springer, 2010.

[30] Nelson, C., and J. Crist, "Predicting Laser Beam Characteristics," *Laser Technik Journal*, Vol. 1, 2012, pp. 36–39.

[31] Rosen, J. H., "Diffraction Patterns Produced by Focused Laser Beams," Technical Report R-925-ARPA, Advanced Research Projects Agency, 1972.

[32] Chen, C. C., "Method for Estimating the Thermal Blooming Effect," Technical Report R-1726-PR, United States Air Force Project RAND, 1975.

[33] Yura, H. T., "Atmospheric Turbulence Induced Laser Beam Spread," *Applied Optics*, Vol. 10, No. 12, 1971, pp. 2771–2773.

[34] Andrews, L. C., and R. L. Phillips, *Laser Beam Propagation through Random Media*, Bellingham, WA: SPIE Press, 2005.

[35] Fante, R. L., "Electromagnetic Beam Propagation in Turbulent Media," in *Proceedings of the IEEE*, Vol. 63, 1975, pp. 1669–1692.

[36] Andrews, L. C., R. L. Phillips, and A. R. Weeks, "Rytov Approximation to the Irradiance Covariance and Variance of a Retroreflected Optical Beam in Atmospheric Turbulence," *Journal of the Optical Society of America A*, Vol. 14, No. 8, 1997, pp. 1938–1948.

[37] Gerson, N. C., "Variations in the Index of Refraction of the Atmosphere," *Geofisica Pura e Applicanta*, Vol. 13, 1948, pp. 88–101.

CHAPTER

6

Contents

6.1 Models for Thresholds

6.2 Single Target and Single Effector

6.3 Multiple Targets: Queueing Theory Approach

6.4 Multiple Sources on a Single Threat

6.5 Number of Effectors to Achieve Minimum Performance

6.6 Concluding Remarks

References

HEL Performance Prediction

This chapter now applies the irradiance functions developed in Chapter 5 to construct measures of HEL DEW performance prediction. The focus throughout this chapter will be on applying (5.11), (5.14), and (5.19), since these irradiance functions capture all the essential factors in laser system power on a target. Hence (5.21) will not be considered further. Due to the fact that these irradiance functions are deterministic it will be of utility to identify an appropriate stochastic phenomenon associated with HEL performance. This will be the initial focus of this chapter, and thereafter some models for performance prediction will be produced. Many of these will be somewhat similar to the HPRF DEW examples, especially when one particular irradiance function is used, namely (5.11). However, the HEL DEW case provides an opportunity to explore the queueing theory approach more widely than in the HPRF DEW case. Additionally, it is possible to answer a number of interesting research problems that arise with HEL DE systems. These will be introduced and addressed toward the end of the chapter.

6.1 Models for Thresholds

Since the HEL DE irradiance functions developed in the previous chapter are deterministic there are not any reference random variables that result in expressions for performance in terms of statistical quantities, in contrast to the HPRF DEW case. Hence it is necessary to identify a stochastic feature of the HEL DEW performance prediction problem. This is by no means an artificial endeavor because there are stochastic features associated with the HEL DE case. Towards this aim, note that from the HEL DEW's perspective, the target threshold level of vulnerability is not a certainty. What can be determined a priori are rough guides to likely threshold vulnerability levels in a given target, as evidenced from the discussion in Section 5.2. Hence the approach introduced in the context of HPRF DE system performance, where the threshold is modeled stochastically, is suitable in the current situation. This will provide a first stochastic element of the performance prediction model. The second stochastic element will be the arrival time of threats into the DEW's theater of operation. As in the HPRF DE case, renewal processes may be applied to model arrivals. However, the first problem to examine is the modeling of the disruption threshold.

Although the previous chapter produced some estimates of irradiance levels on a target to cause a disruptive effect, there is no obvious selection for a model of the disruption threshold. However, several nonempirical based models have been examined for the current study. Given a particular deterministic disruption threshold, one obvious choice is to model it as a uniform random variable, where the known threshold value is assigned to the mean and a suitable variance is introduced to produce an interval matched to the random variable's support. In Bayesian statistical analysis this is equivalent to imposing a uniform prior to match the complete uncertainty concerning a particular variable. This approach was found to result in numerical issues, and hence was the catalyst for searching for a model for thresholds whose probability density function is not discontinuous.

The gamma distribution was then considered, since its mean could be matched to the known disruption threshold and its variance selected so that it decreases as the mean increased. Given most of the disruption thresholds are in the order of powers of tens of W/m^2 this tended to result in unrealistic results in practice due to combinations of products of large and small numbers when evaluating probabilities numerically.

The most useful selection was that of an exponential distribution, whose mean is matched to the known threshold of disruption. This resulted in predicted performance consistent with the results reported in Chapter 5. From an electronic component perspective, the exponential distribution has been used as a model for lifetimes. Hence this provides a practical motivation for its adoption as a threshold model, since component lifetimes and disruption thresholds can be expected to be closely associated. Therefore, throughout this chapter the disruption threshold will be modeled by assuming an exponential distribution with a mean matched to a certain known disruption threshold level. In Chapter 7 the adoption of an exponential distribution for thresholds will be discussed further.

Suppose that the HEL DEW has an irradiance function $I(t)$ and assume that Y is the time it takes for the HEL to defeat a given threat that is assumed to arrive at time $t=0$. Additionally, let the disruption threshold be τ. Then as in the HPRF DEW case,

$$\mathbf{P}(Y>t)=\mathbf{P}\left(\int_0^t I(s)ds<\tau\right) \tag{6.1}$$

is the probability that it takes at least t seconds to defeat the threat. Since τ is being viewed statistically, one may write

$$\mathbf{P}(Y>t)=1-F_\tau\left(\int_0^t I(s)ds\right), \tag{6.2}$$

where F_τ is the distribution function of τ. Since the probability of defeat by time t (viz. $\phi(t)$) is equivalent to $\mathbf{P}(Y\leq t)$ it follows from (6.2) that

$$\phi(t)=\mathbf{P}(Y\leq t)=F_\tau\left(\int_0^t I(s)ds\right). \tag{6.3}$$

The expressions (6.2) and (6.3) are general, and now they are specialized to the particular case of interest.

Under the assumption that τ has an exponential distribution with parameter τ_0^{-1} (so that it has mean τ_0) it follows that

$$\mathbf{P}(Y>t)=e^{-\tau_0^{-1}\int_0^t I(s)ds} \tag{6.4}$$

and

$$\phi(t)=1-e^{-\tau_0^{-1}\int_0^t I(s)ds}. \tag{6.5}$$

The mean of Y may be specified in terms of the irradiance function as follows. Note that the density of Y may be produced by differentiating the complement of (6.2) so that

$$f_Y(t) = \frac{d}{dt}\mathbf{P}(Y < t) = \tau_0^{-1} e^{-\tau_0^{-1} \int_0^t I(s)ds} I(t), \tag{6.6}$$

which has been obtained by the chain rule and the fact that from the fundamental theorem of calculus,

$$\frac{d}{dt} \int_0^t I(s)ds = I(t). \tag{6.7}$$

Consequently, the required mean is

$$\mathbf{E}(Y) = \int_0^{\infty} t\tau_0^{-1} e^{-\tau_0^{-1} \int_0^t I(s)ds} I(t)dt, \tag{6.8}$$

where it is important to note that the density may be zero or undefined after some particular time that must be reflected in the evaluation of (6.8).

An example of the mean service time (6.8) is now considered, based upon the irradiance function (5.19), under the assumption that thermal blooming is negligible. It is supposed that the HEL DEW is facing a single threat, traveling on a straight path toward it, for 11 different speeds (10, 15, 20, 25, 30, 35, 40, 45, 50, 55, and 60 m/s respectively), and that the threat would take 120s to reach the HEL's location if it is not intercepted. The HEL DE operates with a power level of $P_T = 1$ kW and the disruption threshold parameter has been set to $\tau_0 = 10^6$ W/m^2. The environmental propagation parameters have been based upon those in Table 5.1, and so there are three sets of mean service times produced, since there are three sets of wavelength-dependent parameters. Figure 6.1 provides a continuous time interpolation of the mean as a function of the threat speeds. The figure shows the variation of mean service time with wavelength to be small initially and then with a larger deviation as the threat speed increases. The figure also demonstrates that faster traveling threats will have a larger mean service time than slower traveling threats, as expected.

It is now informative to consider the situation where thermal blooming is included, and Figure 6.2 provides an illustration of the complexity in this case. As can be observed, there is a significant increase in the mean service times of threats, especially as the threat speed increases.

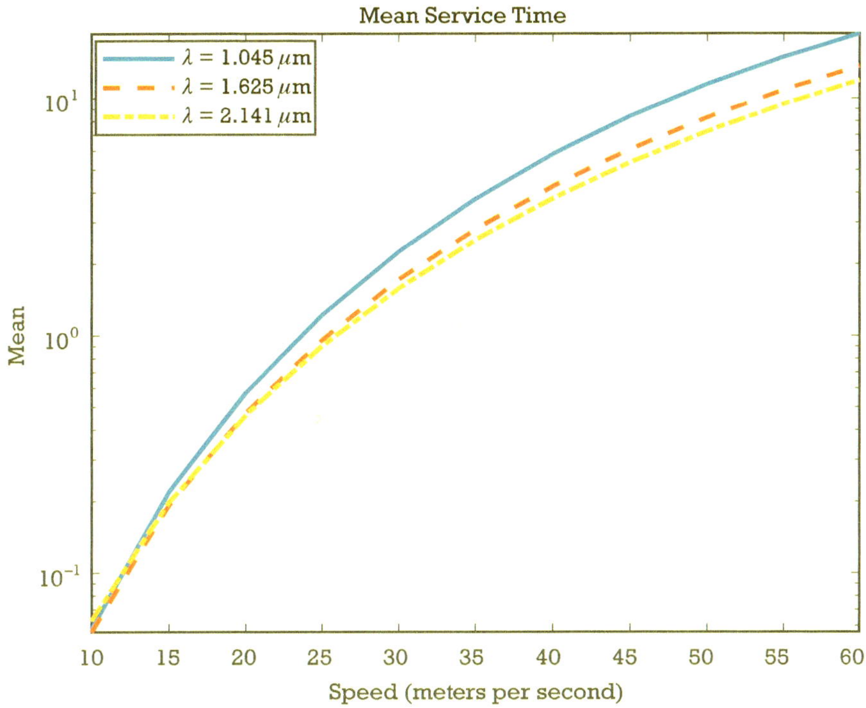

Figure 6.1 A plot of the mean service time of a single threat as a function of its speed, over three different wavelength-dependent cases.

6.2 Single Target and Single Effector

This section provides some examples of a single threat being eliminated by a single HEL DEW. In the first case the irradiance function (5.11) is adopted. This then permits some simplifications in terms of producing expressions for the probability of defeat in closed forms. The approach will be to derive a somewhat general expression for the probability of defeat, and then a tangible example will be considered.

As in the HPRF DE case it is supposed that a threat is traveling toward the DEW's location at a constant speed of v m/s and that it will take T seconds to reach the DEW if not disrupted. Consequently $R(t) = v(T - t)$ for $0 \leq t \leq T$. Define a constant

$$\kappa = \frac{P_T}{2\pi \left[M^4 \delta^2 \frac{\lambda^2}{D^2} + \alpha_J^2 \right]}. \tag{6.9}$$

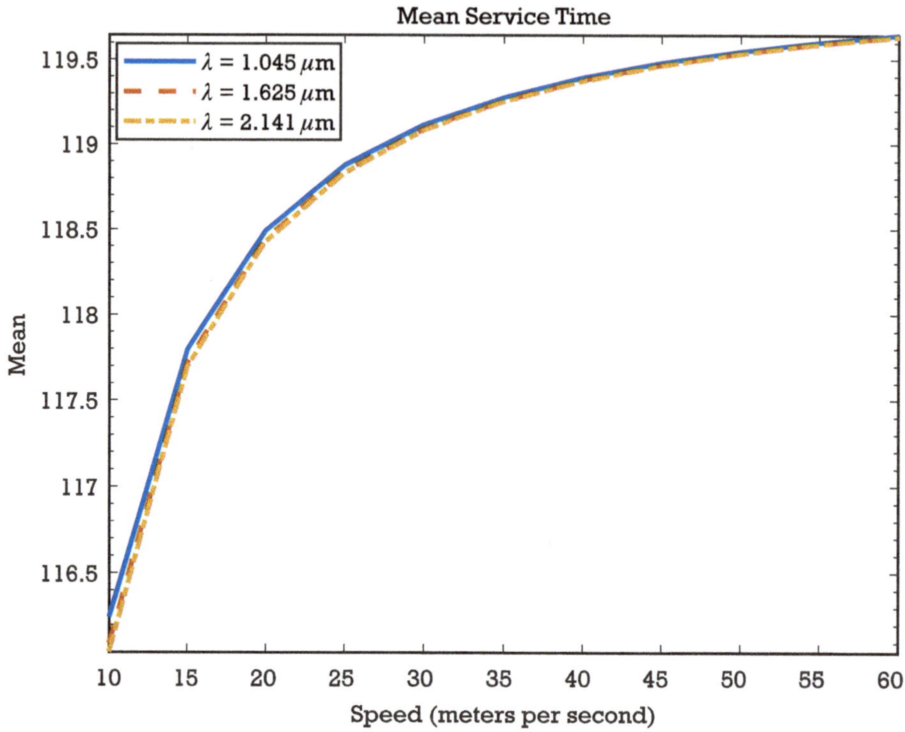

Figure 6.2 Mean service times of a single threat as a function of its speed, over three different wavelength-dependent cases, but in the presence of thermal blooming.

Then a direct calculation establishes that

$$\int_0^t I(s)ds = \frac{\kappa t}{\nu^2 T(T-t)}, \tag{6.10}$$

provided $t < T$ and an application of (6.10) to (6.5) yields

$$\phi(t) = 1 - e^{-\tau_0^{-1} \frac{\kappa t}{\nu^2 T(T-t)}}, \tag{6.11}$$

for $0 \le t \le T$. Hence (6.11) is the probability that the HEL DEW defeats the threat by time t.

Figure 6.3 provides an example of (6.11) for the case where $\nu = 20$ and $T = 120$ with the laser power set to 1 kW ($P_T = 1$ kW). The jitter is assumed negligible in this case, while other parameters have been taken

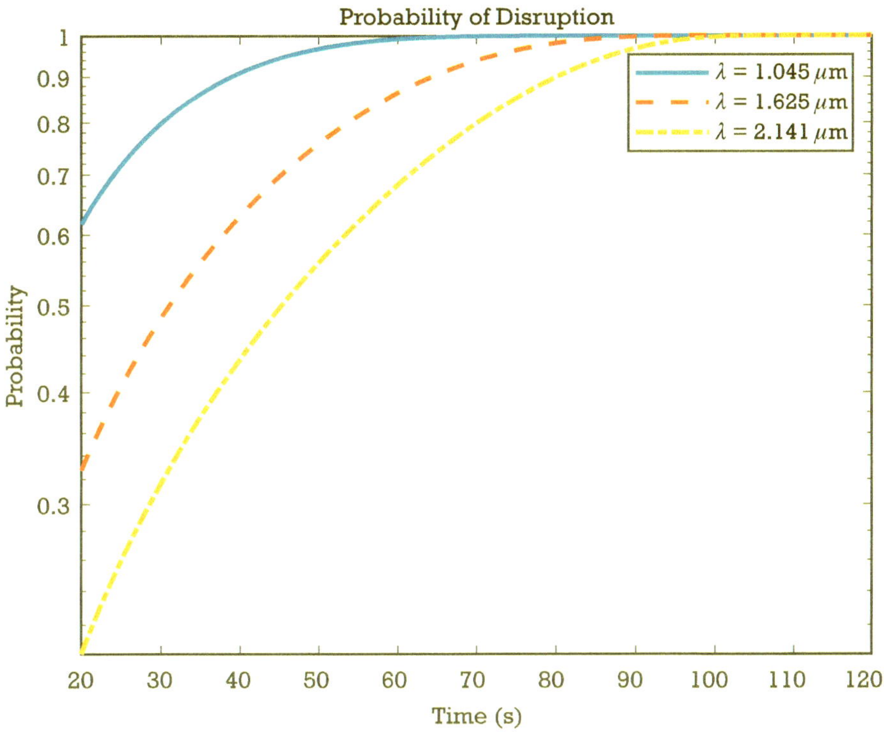

Figure 6.3 An illustration of (6.11) for three wavelength cases.

from Table 5.1. The disruption threshold parameter is $\tau_0 = 10^9$ W/m^2 in this case. The figure shows that as the wavelength increases the likelihood of threat defeat decreases, which is likely due to changes in the environmental factors. Nonetheless the threat is defeated at all wavelengths before 100s. The selection of a somewhat large disruption threshold has been to demonstrate that under these conditions the DEW is likely to produce an irradiance sufficient to disable a number of threats. When this disruption threshold is reduced, the probability of defeat increases significantly across all three wavelength cases such that the DEW is likely to defeat threats on a very short time scale. Note that the threat appears in the combat space at a distance of 2.4 km and, based upon Figure 6.3, a threat with the given disruption threshold is likely defeated almost certainly when it is between 600m and 1 km (wavelength-dependent).

Performance with the irradiance function (5.19) is now examined and numerical methods have been used to approximate (6.5). One observation is that when thermal blooming is negligible the HEL DEW is more likely to defeat threats over a larger range of disruption thresholds. When thermal blooming is included this disruption threshold has a more significant effect if it is larger. Hence the two examples included below are illustrative of only a spectrum of possible performance cases.

In the first example it is supposed that the threat is traveling at $v = 40$ m/s and will take 120s to reach the HEL if not disrupted. The disruption threshold has been set to $\tau_0 = 10^8$ W/m^2 and thermal blooming is assumed negligible, as well as jitter. All other parameters have been selected from Table 5.1, and the HEL DE power is $P_T = 1$ kW. Figure 6.4 plots the estimated probability of disruption, as a function of time, for the three wavelength cases. The figure shows that target defeat is certain after 100s.

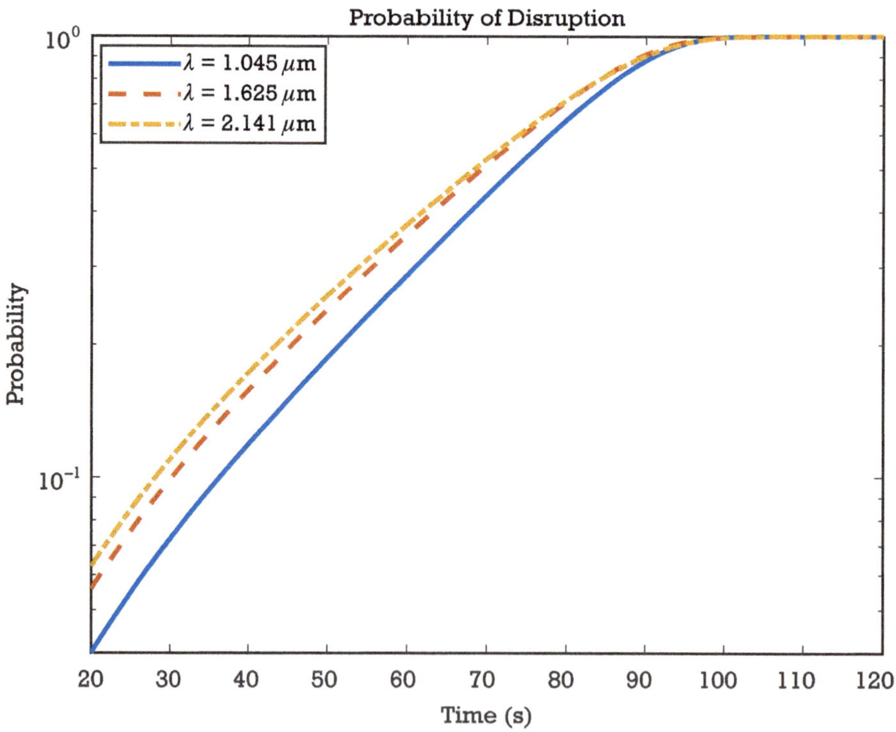

Figure 6.4 Performance prediction using (5.8) for the three wavelength cases.

When thermal blooming is included in the above example, the likelihood of threat defeat is significantly reduced for all three cases, to the point where threat defeat is not likely until the last second before the threat arrives at the DEW. This result is indicative of the inability of the DEW to defeat this threat in the complex propagation environment (figure not included for brevity). At the speed used in this example, there are still issues observed when the disruption threshold is reduced to 10^4 W/m^2. Figure 6.5 is for the case where the disruption threshold has been reduced further to 10^2 W/m^2. A notable feature is that the threat is not likely defeated until the last 5s before it reaches the HEL. When the speed of the threat is changed to 20 m/s it becomes more likely that the DEW defeats it before 110s, and for slower-moving threats this defeat time also reduces.

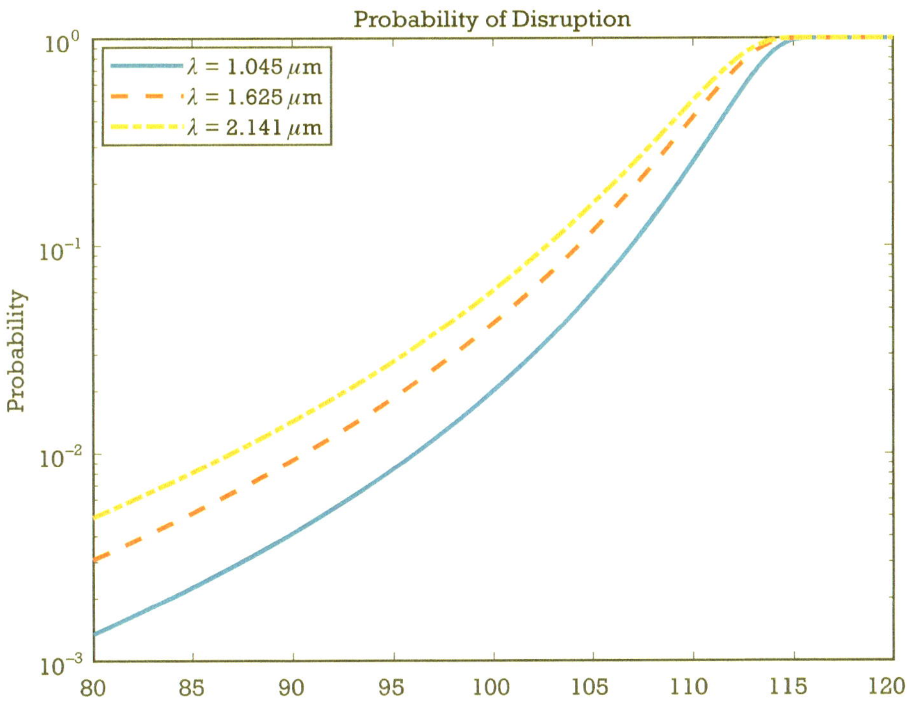

Figure 6.5 Performance prediction using (5.8) for the three wavelength cases in the presence of thermal blooming. The disruption threshold mean is $\tau_0 = 10^2$.

This completes the examination of the single-threat case, and in the next section the somewhat more interesting problem of a HEL DEW defeating a multitude of threats is examined.

6.3 Multiple Targets: Queueing Theory Approach

As in the application of queues in Section 4.2.3 the problem of dealing with multiple threats is now examined, where the queue server is the HEL DEW. The queueing theory model is a very natural construct in which to analyze multiple threats being processed by a single HEL system, since once a threat is detected (or equivalently arrived through the renewal process) it is placed in the system's processing queue and awaits service. During this waiting time the threat still travels toward its intended target, which must be accounted for in the development of the mathematical model.

A preliminary formulation of this queueing theory approach, without reneging, can be found in [1], which is based upon a simpler irradiance function derived from [2]. In the approach in [1] the HEL DEW dwell time is not modeled through consideration of renewal process arrival times and delays of threats in the queue, but instead proposes a somewhat inferior method to account for dwell time, and so does not merit further consideration.

Unlike the development in the HPRF DEW case, the threats do not necessarily have to arrive in a linear pattern. However, it will be assumed for the first example considered in this section that this is the case.

The formulation of the problem is identical to that in Section 4.2.3; it is supposed that threats arrive according to a Poisson process with a given rate, and travel toward the DEW at a constant speed of v_n m/s, and will take T_n seconds to reach the DEW if not disrupted. Hence the distance between the DEW and the nth threat is given by $R_n(t) = v_n(T_N - t)$ for $0 \leq t \leq T_n$. As utilized previously $M(t)$ is the number of threats defeated by time t in the queueing system, where reneging is also permitted.

The probability of disruption is based upon (6.5) with the irradiance function (5.11), since in this case it is possible to specify the service time in terms of arrivals and delays, in a form amenable to simulation. It is important to note that any of the irradiance functions introduced in Chapter 5 can be applied to the queueing theory model.

Based upon analysis of (5.11) it can be shown using the inverse cumulative distribution function simulation approach that the nth service time can be generated through

$$Y_n = \frac{T_n - S_n - D_n}{1 - \frac{\tau_0^{-1}\kappa\nu^{-2}\log(\mathcal{R})}{T_n - S_n - D_n}}, \tag{6.12}$$

where κ is given by (6.9), S_n is the arrival time of the nth threat, and D_n is the corresponding delay the threat experiences in the queue. As specified previously, τ_0 is the threshold parameter and $\mathcal{R}$ is uniformly distributed on the unit interval. The integration utilized to produce (6.12) is analogous to that adopted in (4.25) from the HPRF DEW case in Chapter 4.

Expression (6.12) can be applied to simulate $M(t)$ and its expectation in particular cases. For the example to follow, it is assumed that there are a total of 15 threats, each with $\nu_n = 40$ and $T_n = 120$. The HEL DEW operates as in previous examples with model parameterization based upon Table 5.1, except only the wavelength of $\lambda = 1.045$ μm is investigated. The HEL operates with a power of $P_T = 1$ kW and jitter is assumed negligible. Finally the disruption threshold parameter has been set to $\tau_0 = 10^6$ W/m^2. In order to evaluate probabilities associated with $M(t)$, a Monte Carlo simulation with 10^5 runs has been used.

In the first example the rate of arrivals and threat reneging times are assumed to be unity. Figure 6.6 plots three complementary distribution functions for this case, showing that by 30s all the threats are likely defeated by the HEL DEW. When the arrival rate is tripled it is not until 80s that all threats are likely eliminated.

Other variations, as in the HPRF case, may be considered, including the construction of the mean number of threats defeated. For the problem parameterization used to produce Figure 6.6, the mean number of threats defeated is plotted in Figure 6.7, which provides a useful synthesis of the plot in Figure 6.6.

For a second example it is supposed that two HEL DEWs are providing protection to an asset being targeted by a series of threats, a scenario depicted in Figure 6.8. It is assumed that the series of threats is directed toward O in this figure (the origin), and that the HEL DE defensive positions are located at positions B_1 and B_2 as shown. These are fixed during the engagement for convenience, and are located at prescribed distances (d_1 and d_2) from O, and at angles ϕ and θ to the y-axis as illustrated. As depicted in Figure 6.8, threat j is denoted $V_j(t)$ where the time-dependence is to reinforce the

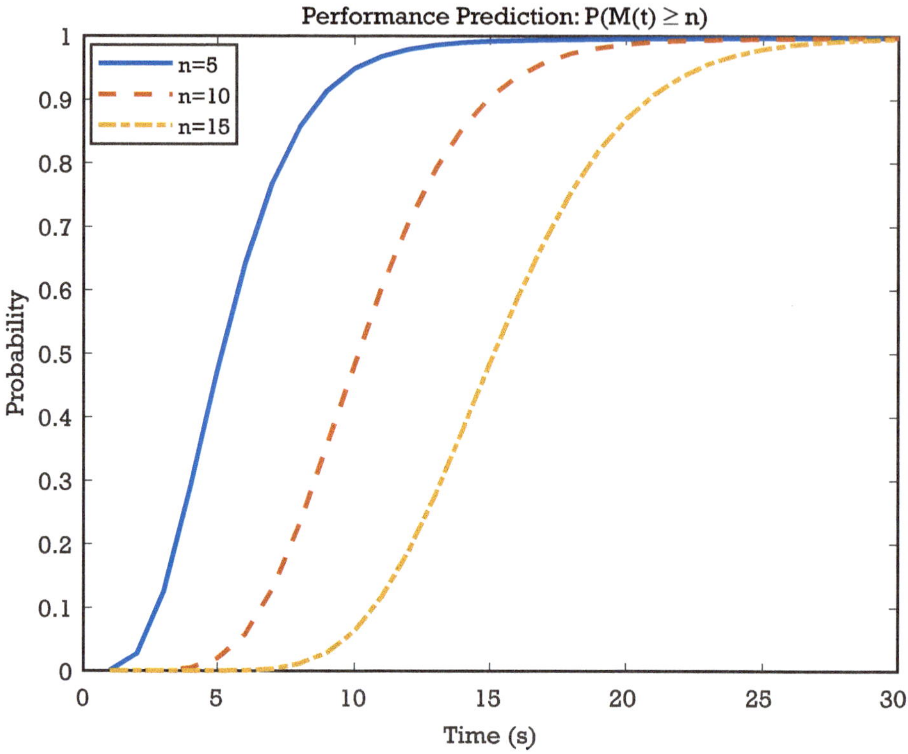

Figure 6.6 Example of processing multiple threats through a queueing theory interpretation of the threat defeat problem.

fact that the threats are not stationary. The distance between threat j and effector B_1 at time t is denoted $R_{(j,1)}(t)$, and similarly $R_{(i,2)}(t)$ is the distance between threat i and B_2 at time t. The scenario under consideration can be modeled by a network of two queues, with each queue accounting for the performance for the HEL DEW it is representing. It will be supposed that the threats are generated through a Poisson process, and once they have arrived into the combat scene they are assigned alternately between the two queues. In particular the assignment algorithm is:

$$V_j(t) \text{ assigned to the queue corresponding to effector } \begin{cases} B_1 & \text{if } j \text{ is even;} \\ B_2 & \text{if } j \text{ is odd.} \end{cases}$$

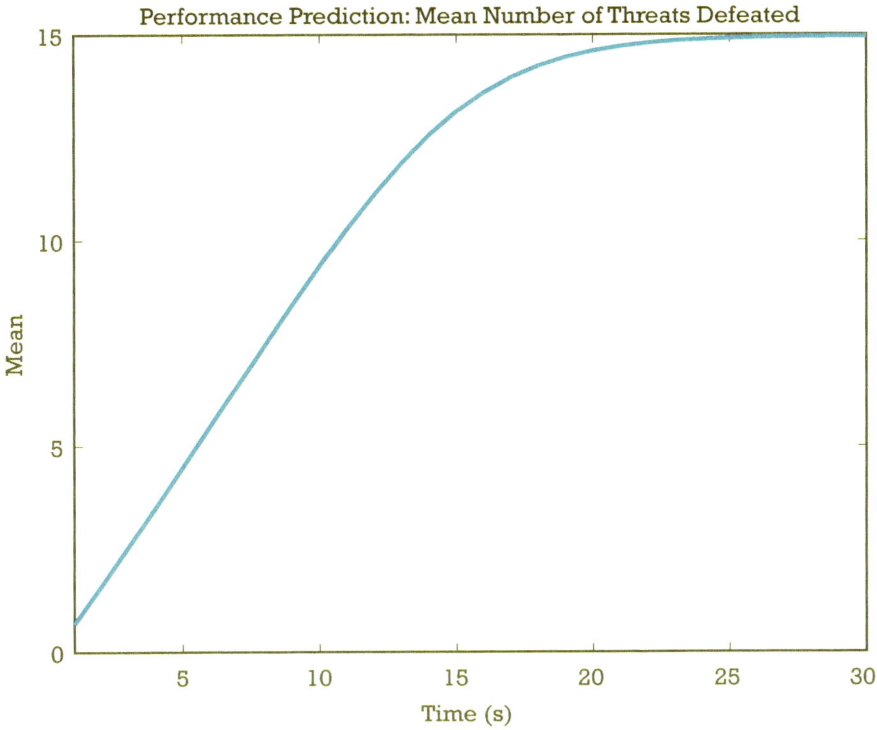

Figure 6.7 Mean number of threats defeated in the HEL DEW example.

This assignment algorithm has been selected arbitrarily; other variations are permissible and will be discussed further in Chapter 7 in terms of useful future research problems.

As in the previous queueing example considered in this chapter, it will be supposed that threats may renege but once assigned to a particular queue it remains in that queue until its exit. Suppose that $M_1(t)$ is the number of threats that have been processed by the first queue (corresponding to HEL B_1), and let $M_2(t)$ be defined similarly for the second queue, and corresponding to HEL B_2. Then the total number of threats defeated by time t is $M_1(t) + M_2(t)$. It will also be assumed that the queues, and so the HELs, operate independently.

It is not difficult to analyze the performance for any of the two queues separately; this can be done as outlined in the previous example based upon

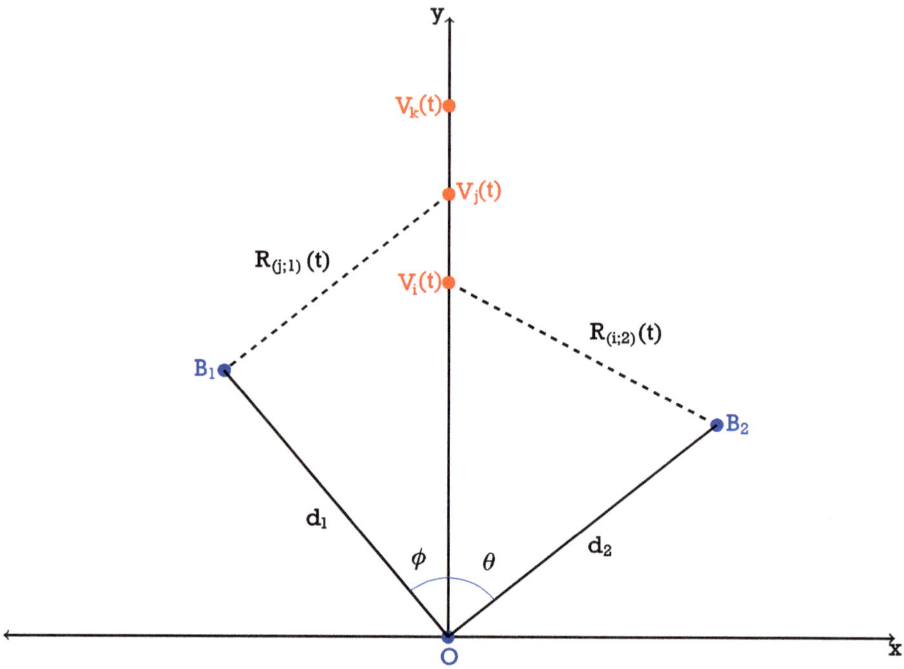

Figure 6.8 The geometry of the engagement scene. Here there are two HEL DEWs operating independently in terms of processing assigned threats, but when the latter arrive through the renewal process, an algorithm will assign threats to either of the queues. All threats $V_i(t)$, $V_j(t)$, and $V_k(t)$, are directed toward O and the HEL DEWs are B_1 and B_2, respectively.

the results included in Section 4.2.3. To analyze the situation from a network of queues perspective, one may analyze performance based upon the mean, since the mean is additive and so the mean number of threats defeated by time t is the sum of the mean number of threats defeated in each individual queue.

Since the two queues function independently it is also possible to specify the complementary distribution function of the number of threats defeated by time t. Assume that a maximum of N threats are to be processed by the first queue. Then by conditional probability, and for any $n \in \mathbf{N}$,

$$\mathbf{P}(M_1(t) + M_2(t) \geq n) = \sum_{k=0}^{N} \mathbf{P}(M_1(t) = k)\mathbf{P}(M_2(t) \geq n - k)$$

$$= \sum_{k=0}^{N} [\mathbf{P}(M_1(t) \geq k) - \mathbf{P}(M_1(t) \geq k+1)]$$
$$\times \mathbf{P}(M_2(t) \geq n-k), \tag{6.13}$$

where the second equality has been specified completely in terms of complementary distribution functions to facilitate numerical evaluation. It is important to observe that when evaluating (6.13) many of the complementary distribution function probabilities may in fact take values either of zero or one.

It is now necessary to introduce parameterizations in order to generate tangible results. With reference to Figure 6.8 suppose that $\phi = \frac{\pi}{4}$ and $\theta = \frac{\pi}{3}$. It will be assumed that the threats are homogeneous; hence threat j travels at a speed of ν m/s and would take T seconds to reach its target if not intercepted. Then by an application of the cosine rule it follows that

$$R^2_{(j,1)}(t) = \nu^2 \left[\left(t - T + \frac{d_1}{\sqrt{2}\nu_j} \right)^2 + \frac{d_1^2}{2\nu^2} \right]. \tag{6.14}$$

In order to facilitate the analysis it will be supposed that $d_1 = \sqrt{2}\nu T$ so that (6.14) reduces to

$$R^2_{(j,1)}(t) = \nu^2 \left[t^2 + T^2 \right]. \tag{6.15}$$

Assume that the disruption probability is based upon (6.5) with the irradiance function (5.11), as in the previous example. Then

$$I_j(s) = \frac{\kappa}{\nu^2[s^2 + T^2]}, \tag{6.16}$$

where κ is given by (6.9) and it is required to integrate (6.16) over the observational window $[S_j + D_j, S_j + D_j + t]$ in order to evaluate (6.5) in this case. The integral can be evaluated through appeal to the well-known integral

$$\int_0^t \frac{ds}{s^2 + a^2} = \frac{1}{a} \arctan\left(\frac{t}{a}\right), \tag{6.17}$$

as well as the identity

$$\arctan(x) - \arctan(y) = \arctan\left(\frac{x - y}{1 + xy}\right), \tag{6.18}$$

results that can be sourced from any elementary calculus textbook such as [3]. By applying (6.17) and (6.18) one can demonstrate that

$$\int_{S_j+D_j}^{S_j+D_j+t} I_j(s)ds = \frac{\kappa}{\nu^2 T} \arctan\left(\frac{tT}{T^2+(S_j+D_j)(S_j+D_j+t)}\right). \quad (6.19)$$

Expression (6.19) may now be applied to a result equivalent to (6.4) to yield the complementary distribution function

$$\mathbf{P}(Y_j > t) = e^{-\tau_0^{-1}\frac{\kappa}{\nu^2 T}\arctan\left(\frac{tT}{T^2+(S_j+D_j)(S_j+D_j+t)}\right)}, \quad (6.20)$$

which may then be used to simulate the service time of the jth threat. Using the inverse cumulative distribution function method applied to (6.20) it can be shown that Y_j may be simulated via

$$Y_j = \Psi\frac{\left[T^2+(S_j+D_j)^2\right]}{T-\Psi(S_j+T)}, \quad (6.21)$$

where

$$\Psi = -\tan\left(\frac{\tau_o \log(\mathcal{R})\nu^2 T}{\kappa}\right), \quad (6.22)$$

and $\mathcal{R}$ is a uniformly distributed random variable on the unit interval.

For the second HEL DEW a similar set of calculations can be used to show that

$$R^2_{(j,2)}(t) = \nu^2\left[t^2+3T^2\right], \quad (6.23)$$

which has been derived under the assumption that $d_2 = 2\nu T$, and consequently the service time of the threat in the second queue is given by

$$Y_j = \Psi\frac{\left[3T^2+(S_j+D_j)^2\right]}{\sqrt{3}T-\Psi(S_j+T)}, \quad (6.24)$$

where

$$\Psi = -\tan\left(\frac{\tau_0 \log(\mathcal{R})\nu^2\sqrt{3}T}{\kappa}\right), \quad (6.25)$$

and as before $\mathcal{R}$ is a uniformly distributed random variable on the unit interval.

Hence (6.21) and (6.24) may be used to simulate the service times of threats in the respective queues. Specific numerical values are now adopted to generate performance results. Toward this aim, suppose that $\nu = 40$ and

$T = 120$, with the first HEL operating with a power level of 10^3 W, while the second with a power level of 10^4 W. The laser parameterizations are as used previously, so that $D = 0.8$, $M^2 = 1$, $\lambda = 1.045$ μm, and jitter is assumed negligible. The threshold parameter has been selected to be $\tau_0 = 10^6$ W/m^2 throughout. A total of 16 threats were generated, so that eight were assigned to the first HEL and eight to the second, and Monte Carlo simulations with 10^5 runs were used to estimate the relevant probabilities. The first example assumes that the arrival rate and the reneging rate is unity. Figure 6.9 plots the complementary distribution function $\mathbf{P}(M_1(t) + M_2(t) \geq n)$ for $n \in \{1, 8, 16\}$. Figure 6.10 plots the mean number defeated for each respective queue, as well as for the system considered as a network of two queues. These two figures show that the first HEL DEW tends to perform slightly

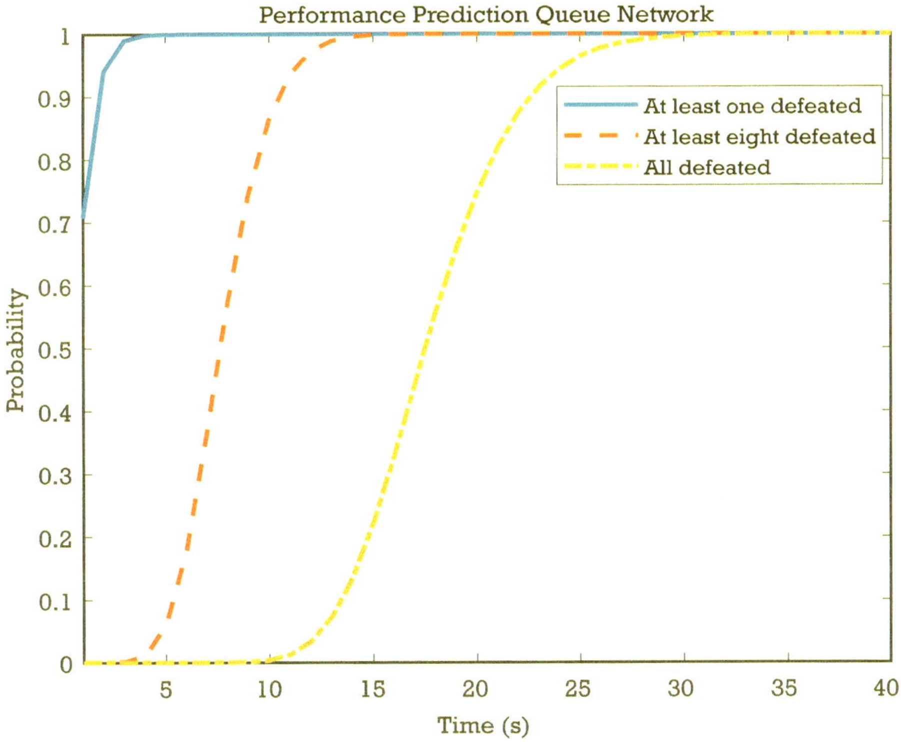

Figure 6.9 Complementary distribution function $\mathrm{P}(M_1(t) + M_2(t) \geq n)$ for $n \in \{1, 8, 16\}$.

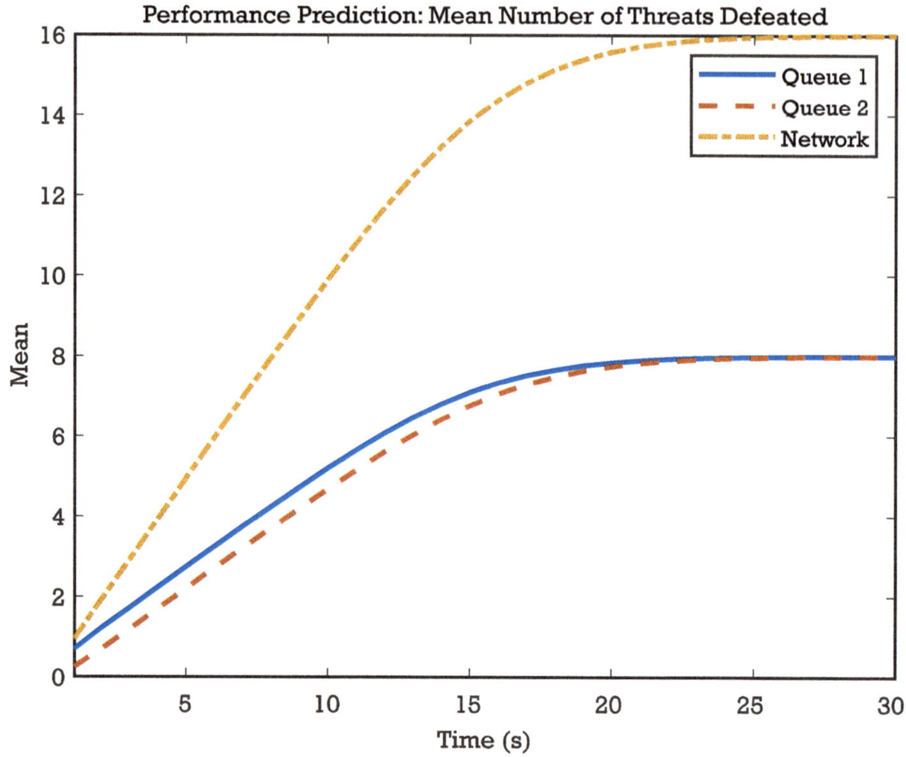

Figure 6.10 Means of each individual queue and the network of two queues.

better (an attribute of the assumed geometry) but as a whole the network is likely to defeat all threats after 30s since the engagement was initiated.

It is interesting to examine the same situation when the arrival rate is increased to 4; Figure 6.11 plots the complementary distribution functions while Figure 6.12 shows the mean functions. These figures show that the faster arrival rate results in it taking much longer to defeat the threats at the various tier levels.

As a final remark on the queueing theory approach considered in this section, there is an interesting variation of the single server queue problem that is discussed in Appendix C. This is concerned with the distribution of delays in a queueing system and its usefulness in determining whether a HEL DEW can be defeated if the threat model characteristics are varied.

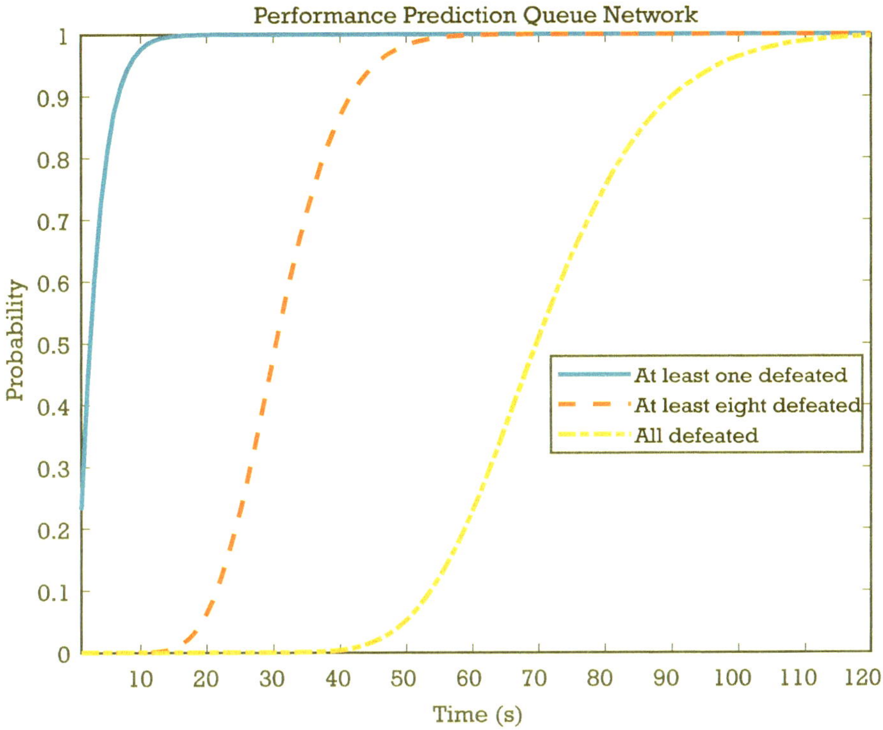

Figure 6.11 Complementary distribution function for the network of queues when the arrival rate is four times that of Figure 6.9.

6.4 Multiple Sources on a Single Threat

In all the previous examples a particular HEL DEW was applied to disrupt a specific threat. The purpose of this section is to show how the scenario in which a series of HEL DEWs are targeting a single common threat can be addressed. Consider the situation illustrated in Figure 6.13 where a series of HEL DEWs, denoted $B_1, B_2, \ldots, B_N$, are facing a single threat V. In this scene it is assumed that the threat is traveling along the y-axis toward the origin, and that the HEL DEWs are located at fixed positions as shown. Suppose that each of these HEL DEs target the threat at the same time, but strike different threat attributes. This assumption is to avoid the complexities of beam interactions on the target surface. The interest is in producing a single measure of the likelihood of threat defeat under such a situation.

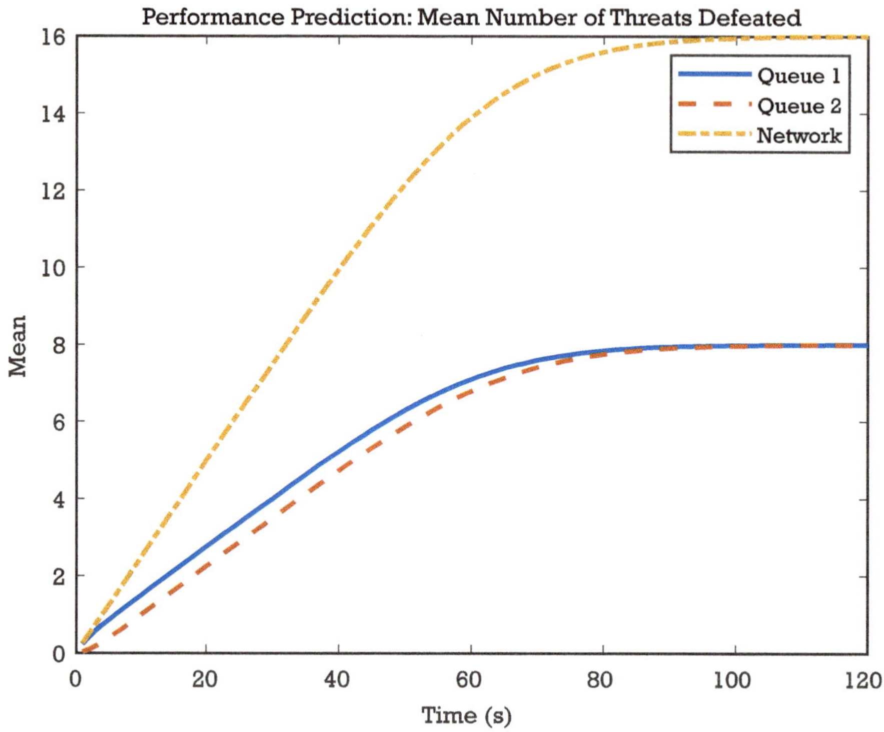

Figure 6.12 Expected number of threats defeated with an arrival rate of four.

Suppose that $\phi_j(t)$ is the probability that HEL DEW j defeats the common threat by time t. Then the probability that the threat is defeated by time t is the probability that at least one of the HEL DEWs defeat the target by time t. Assuming that each HEL DEW operates independently it can be shown that

$$\mathbf{P}(\text{Threat defeated by time } t) = 1 - \prod_{j=1}^{N}(1 - \phi_j(t)), \tag{6.26}$$

which has been based upon the fact that the desired probability is easier to compute under negation of events. This expression is analogous to that developed in (4.6).

An important observation concerning (6.26) is that it is not necessary to assume that each DEW's beam will have a common disruption threshold

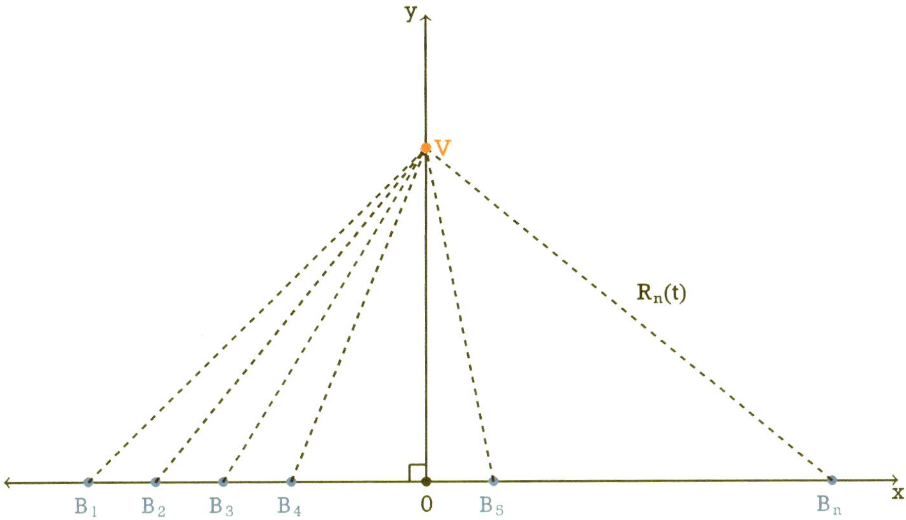

Figure 6.13 An illustration of the combat scene where a series of HEL DEWs, denoted $B_1, B_2, \ldots, B_n$, are focused on a common threat (V).

on the target. Given that the series of DEWs are targeting different patches of surface area on the threat one may suppose that DEW j will require an irradiance on the target of τ_j W/m^2 in order to cause a disruption. However, in the particular example to be considered it will be assumed that the threat has a homogeneous surface area vulnerability.

To illustrate (6.26) it is necessary to introduce model parameterizations. Hence suppose that the threat is traveling toward the origin at a speed of ν m/s, taking T seconds to reach the origin if not disrupted. Assume that the HEL DEWs are spaced at equal intervals so that the nth HEL is at a distance of n meters from the origin. Then based upon the Pythagorean theorem it follows that for the nth HEL the distance to the threat is $R_n(t) = \sqrt{n^2 + \nu^2(T-t)^2}$. The probability of disruption (6.5) is applied, together with (5.11), as utilized previously in this chapter. Therefore define

$$\phi_n(t) = 1 - e^{-\tau_0^{-1} \int_0^t I_n(s)ds} \tag{6.27}$$

where

$$I_n(s) = \frac{\kappa}{R_n^2(s)}, \tag{6.28}$$

and κ is defined via (6.9). With the particular form of $R_n(t)$ defined above it is possible to evaluate (6.28) using the identity[1]

$$\int_0^t \frac{1}{(x+a)^2+b^2} = \frac{1}{b}\arctan\left(\frac{tb}{b^2+a(t+a)}\right) \tag{6.29}$$

to produce

$$\phi_n(t) = 1 - e^{-\tau_0^{-1}\frac{\kappa v - 1}{n}\arctan\left(\frac{ntv}{n^2+T_n v_n^2(T_n - t)}\right)}. \tag{6.30}$$

Consequently (6.30) can be applied to (6.26) to provide an expression for the defeat of the threat when N HEL DEWs are available.

To illustrate these results a specific example is now investigated, where it is assumed that $N = 10$ homogeneous HEL DEWs are available, operating under conditions as utilized in the previous section. The laser power is assumed to be $P_T = 1$ kW, operating at a wavelength of 1.045 μm and $\tau_0 = 10^9$ W/m^2. Jitter is assumed negligible and the threat is supposed to travel at 40 m/s with $T = 120$. Figure 6.14 plots the likelihood of threat defeat, showing it for the nearest neighbor ($n = 1$) and the combined team. As can be observed the combined team is more likely to defeat the threat earlier than the single HEL DEW.

6.5 Number of Effectors to Achieve Minimum Performance

The development in the preceding section can also be utilized to answer a slightly different question, namely that of how many HEL DEWs would be required to ensure a minimum likelihood of DEW team defeat of a single threat. Suppose that one would like to ensure that this latter minimum performance level is $\epsilon \in (0, 1)$. The time-domain where this minimum level can be achieved is of considerable interest.

The following analysis has appeared in [4] where more extensive examples of it may be found, and it also utilizes the irradiance function discussed in [1]. The examples considered here are based upon results derived in this chapter as well as Chapter 5.

1. See, for example, [3].

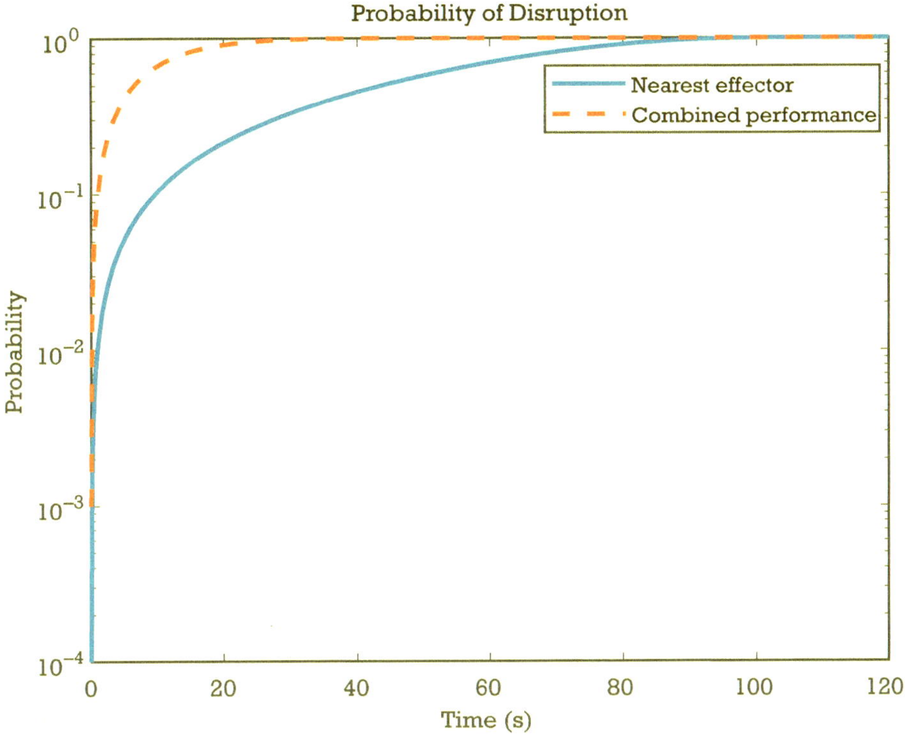

Figure 6.14 A comparison of a single HEL DEW effector's performance and that of a set of 10 HEL DEWs operating independently to defeat the threat.

In order to answer the question of interest, in view of (6.26) what is required is to determine the smallest N so that

$$1 - \prod_{j=1}^{N} (1 - \phi_j(t)) \geq \epsilon. \tag{6.31}$$

By rearranging (6.31) and applying properties of logarithmic functions it can be shown that (6.31) is equivalent to

$$\sum_{j=1}^{N} \frac{\log(1 - \phi_j(t))}{\log(1 - \epsilon)} \geq 1. \tag{6.32}$$

Consequently, in view of (6.32), what is required is to determine the solution to

$$\min_{N}\left\{\sum_{j=1}^{N}\frac{\log(1-\phi_j(t))}{\log(1-\epsilon)}\geq 1\right\}. \tag{6.33}$$

It is worth observing that the selection of N may not necessarily be independent of t, as the following examples will demonstrate.

The example is again based upon the formulation adopted previously, and in particular (6.30) is applied here, so that it is being supposed that a series of HEL DEWs are positioned as in the formulation in the preceding section, with exactly the same HEL DEW operating characteristics.

The following figures plot (6.32) as a function of t, for a series of values for N, and with the requirement that $\epsilon = 0.95$. Figure 6.15 plots this expression

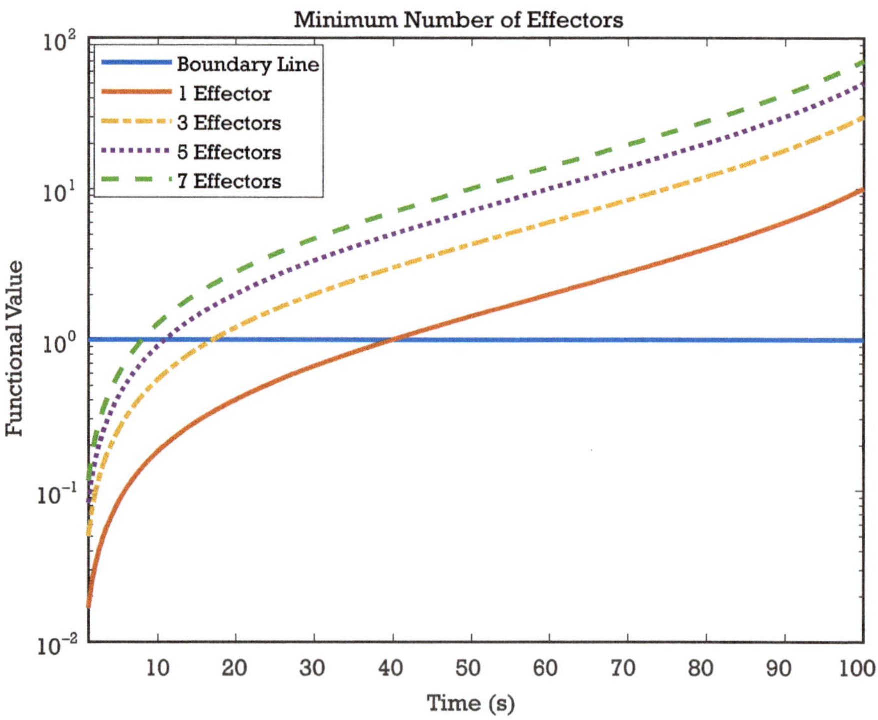

Figure 6.15 Number of effectors to achieve a performance level of at least $\epsilon = 0.95$, illustrated when $N \in \{1, 3, 5, 7\}$.

for $N \in \{1, 3, 5, 7\}$ while Figure 6.16 corresponds to $N \in \{10, 12, 15, 20\}$. From these figures one can conclude that for the desired level of disruptive effect it will take almost 20 HEL DEWs to guarantee the desired level uniformly over time. However, for practical purposes the number of effectors required will be time-dependent relative to how far the threat has traveled. Hence in view of Figure 6.15 after 40s, one HEL will guarantee the desired level of performance. If it was desired to ensure the minimum probability of disruption is achieved earlier, then after roughly seven seconds it would take only seven effectors to guarantee the minimum, while at roughly six seconds it would take ten effectors to achieve the same minimum level of performance.

6.6 Concluding Remarks

This chapter completes the examination of DE performance prediction models, and in particular, proposed a way in which performance could be assessed

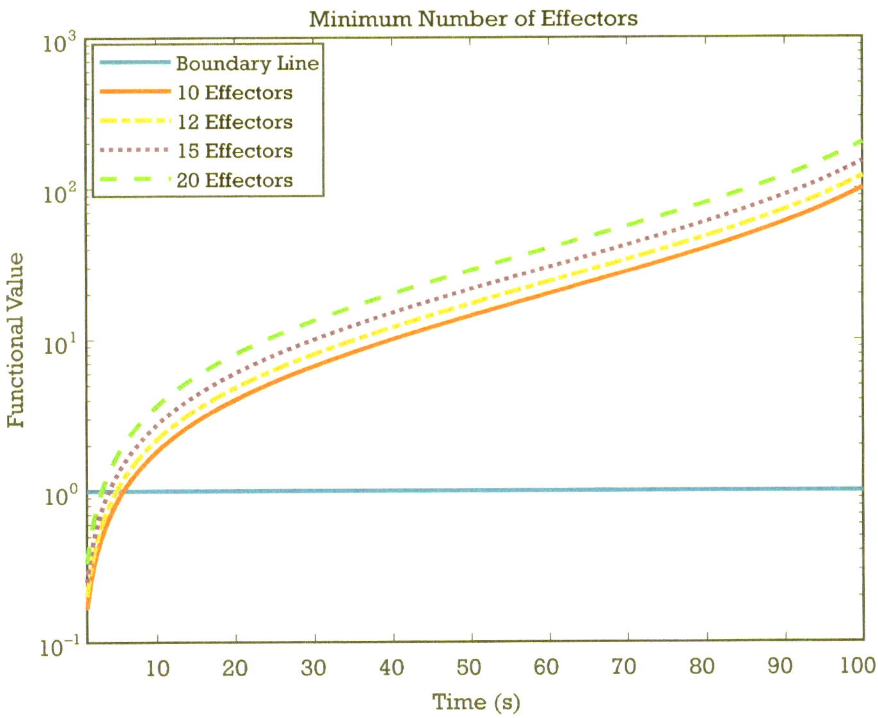

Figure 6.16 A similar plot to that in Figure 6.15, except $N \in \{10, 12, 15, 20\}$.

for HEL DEWs using the irradiance functions derived in Chapter 5. The complexity of turbulence, and in particular thermal blooming, was demonstrated. The queueing theory approach naturally applied in the context of HEL DEW performance and it was shown how the results in Chapter 2 could be applied. Additionally, a two-queue network example was investigated to model the case where two HEL DEWs are defending an asset. Finally, the chapter examined how to model the case where a number of HEL DEWs are simultaneously attempting to disrupt a target, and also demonstrated how the minimum number of HEL DEWs could be determined, to ensure a particular probability of disruption.

The next chapter will outline a number of open problems that have arisen as a result of this chapter's analysis, as well as some that emerged from the HPRF DEW case.

References

[1] Weinberg, G. V., "Performance Prediction of Directed Energy Weapons," *Progress in Electromagnetics Research M*, Vol. 108, 2022, pp. 79–88.

[2] Sprangle, P., J. Penano, and B. Hafizi, "Beam Combining and Atmospheric Propagation of High Power Lasers," Technical report, Naval Research Laboratory, Washington D.C., 2011.

[3] Thomas, G. B., *Calculus and Analytic Geometry*, Reading, U.K.: Addison-Wesley, 1965.

[4] Weinberg, G. V., "Guarantees of Minimum Performance Levels with Directed Energy Weapons," *Progress in Electromagnetics Research Letters*, Vol. 105, 2022, pp. 27–32.

CHAPTER

7

Contents

7.1 HPRF DEW Considerations

7.2 Modeling Synchronized DEW Systems

7.3 Active Protection Systems

7.4 Adaptive Optics

7.5 Validation of Performance Models

7.6 Development of Performance Prediction for Acoustic Systems

References

Future Research Directions

This book has focused on the mathematical modeling and predictive analysis of the performance of DE systems, with a particular focus on DEWs. It has been written as a foundational study of this subject area. Hence there is much more potential research work available within the problems examined in this analysis. Therefore it is the objective of this final chapter to outline further avenues of research available to the interested practitioner.

7.1 HPRF DEW Considerations

There are several avenues of research available in the context of HPRF DE system performance prediction. Recall that in Chapter 4 only one effector was considered throughout. It would be of interest to examine the case where a number of HPRF DEWs are operating simultaneously, which requires consideration of the potential interaction of waveforms and their effects on targets. It may be of interest to suppose that a set of narrowband effectors are operating at different power levels and at different wavelengths, and examine the resultant probability of disruption, assuming that a given threat, or series of threats, have wavelength-dependent disruption thresholds. This could be combined with any of the multiple-threat models introduced in Chapter 4.

Another useful area of investigation would be to extend the wideband performance prediction development introduced in Chapter 4, which was based upon the spectral analysis discussed in Chapter 3. There is merit in examining other wideband signals and investigating whether the proposed approach in this book has further applicability.

7.2 Modeling Synchronized DEW Systems

The network of two queues example investigated in the latter part of Section 6.3 assumed that the HEL DEWs operated independently. However, it would not only be of interest to extend the analysis to multiple HEL DEWs providing regional defense but to introduce an engagement protocol that determines when a particular HEL is assigned to a specific threat. Hence, based upon a target assignment or allocation algorithm, threats that have arrived into the DEW's theater of operation could be assigned to a particular queue associated with a given HEL. This allocation may be based upon the nearest available DEW to the threat once it has arrived, or may also be used to divert surplus threats away from HELs with a significant backlog of queued threats. There has been much work on weapon target assignment optimization that may facilitate determining a suitable way in which to synchronize HEL DEWs [1].

Another useful feature that could be incorporated into this network of queues framework is allowing threats to be reassigned between queues. Since a HEL will experience delays between processing threats it may be useful if the network was able to pass threats onto other HEL DEW processing queues, especially if there were queues that had few customers. The mathematical framework in which such a requirement may be introduced is known as an open migration process [2], and under arrivals from a Poisson process there is considerable literature examining the mathematical modeling of such queueing systems [3–5]. The issue in the current context is that the service time random variable is clearly not exponentially distributed, which is a fundamental assumption used in much of the literature. Hence there are several open problems available in this context. Firstly developing an open migration process architecture with Poisson arrivals, but with service times generated through the HEL DEW process, is one possible avenue of research. A second is to examine conditions under which an exponential approximation of the HEL DEW defeat time may be valid.

Nonetheless, developing a performance prediction capability when threat scheduling and processing is modeled through an open migration process will be a very interesting and useful future research exercise.

7.3 Active Protection Systems

Active protection systems (APS) provide a suitable application for the HEL DEW performance prediction work developed in Section 6.3. An APS is designed to provide an AFV with an automatic threat detection, tracking, and disruption capability [6]. HPRF and HEL DEWs are therefore potential candidates for this disruptive feature. Collaborative APS (CAPS) is an emerging technological concept that is concerned with the case where a series of APS are integrated to work collaboratively to defeat threats that a team of AFVs may face, which also falls under the synchronization of DEW systems discussed in the preceding section. A key difference in this case is the requirement that an automatic threat detection and tracking capability is required and must be integrated into the direct fires action of the APS.

The analysis in this book has proceeded under the assumption that threat arrival evolves according to a renewal process, which can be used to generate the arrival times of threats into the combat space. This has been utilized to bypass detection of threats. In the case of CAPS it would be more interesting to introduce sensor-based results directly into the analysis. This has been done in a number of preliminary studies of CAPS but has been based upon a low-resolution sensor model for an X-band radar [7, 8]. Given a laser-based system may be used for detection (lidar being the prime example) it may also be useful to consider a HEL system with a shared aperture for both sensing and effecting, and incorporate this into a generic model for CAPS with suitable scheduling algorithms for threats, as discussed in the preceding section. This would be a modeling development of considerable advantage to those investigating AFV team defense with CAPS utilizing emerging technology.

7.4 Adaptive Optics

As remarked in Chapter 5, and further examined briefly in Appendix B, adaptive optics is a way in which the effects of turbulence may be minimized. It would therefore be of significant utility to develop some model parameterizations to allow the performance analysis of HEL DE systems with this feature. This could then be used to produce an irradiance function that subsequently could be applied to the examples in either of the two preceding sections. A preliminary version of this has been included in Appendix B, which may be used as a basis for this development.

7.5 Validation of Performance Models

The models for DEW performance introduced in this book have not been validated against real data, and so it would be of relevance to undertake some form of model validation. Another important part of the HEL DEW model development is the statistical models for the threshold, adopted for the performance prediction in Chapter 6. As discussed, the chosen exponential model was somewhat arbitrary, and so there is merit in the examination of disruption threshold measurements, for various features of targets, to determine an empirically-generated statistical model for thresholds.

7.6 Development of Performance Prediction for Acoustic Systems

It was suggested in Chapter 1 that the principles of performance prediction developed in this book may be applied to quantify the effectiveness of acoustic-based weapon systems such as the LRAD. There are well-known expressions for the acoustic intensity produced by a sound-source directed towards a target, which can be found in books examining sonar performance such as [9–11]. The results in the latter books may be used to produce associated intensity functions for operation in the air, as opposed to undersea applications. It would be of interest to apply the approaches in this book to develop an acoustic weapon performance prediction capability.

References

[1] Hocaoğlu, M. F., "Weapon Target Assignment Optimisation for Land-Based Multi-Air Defense Systems: A Global Programming Approach," *Computers and Industrial Engineering*, Vol. 128, 2019, pp. 681–689.

[2] Whittle, P., "Equilibrium Distributions for An Open Migration Process," *Journal of Applied Probability*, Vol. 5, 1968, pp. 567–571.

[3] Daley, D. J., "Queueing Output Processes," *Advances in Applied Probability*, Vol. 8, 1976, pp. 395–415.

[4] Kelly, F. P., "Networks of Queues with Customers of Different Types," *Journal of Applied Probability*, Vol. 12, 1975, pp. 542–554.

[5] Kelly, F. P., "Networks of Queues," *Advances in Applied Probability*, Vol. 8, 1976, pp. 416–432.

[6] Yang, L., and J. Xu, "Analysis on the Development of Active Protection System for Tanks and Armored Vehicles," *Journal of Physics: Conference Series, (1855)*, 2021.

[7] Weinberg, G. V., "Quantification of Combat Team Survivability with High-Power RF Directed Energy Weapons," *Progress in Electromagnetics Research M*, Vol. 102, 2021, pp. 1–11.

[8] Weinberg, G. V., "A Queueing Theoretic Approach for Performance Prediction of Collaborative Active Protection Systems," in *24th International Conference on Modelling and Simulation*, 2021, pp. 904–910.

[9] Urick, R. J., *Principles of Underwater Sound*, New York: McGraw-Hill, 1983.

[10] Ainslie, M. A., *Principles of Sonar Performance Modeling*, New York: Springer, 2010.

[11] Rossing, T. D., *Handbook of Acoustics*, New York: Springer, 2007.

APPENDIX

A

Contents

A.1 Airborne Threats
A.2 Ground-Based Threats
A.3 Sea-Based Threats
References

Emerging Threat Exemplars

One of the primary functions of DE systems is to counter emerging threats that are now appearing in prototype form and will likely appear integrated into the future fighting forces. Hence this appendix provides a brief survey of technology that is likely to appear in the near future, or systems that have been already deployed. As such, the focus is exclusively on uncrewed systems, and exemplars across the three domains of air, land, and sea will be provided.

A.1 Airborne Threats

From an airborne perspective UAVs are an obvious threat, and these come in a wide range of sizes for different applications. These vary from micro-UAVs, such as the PD-100 Black Hornet Nano UAV [1], shown with Australian Army troops in Figure A.1, to full-scale ISR drones such as the Northrop Grumman MQ-4C Triton [2], shown in Figure A.2, which is currently being introduced into operations with the Royal Australian Air Force.

There is also a wide range of COTS UAVs available. As an example DJI's Mavic 2 Air is

Figure A.1 An example of a micro-UAV, being trialed by the Australian Army. (Source: Australian Army.)

a quadcopter with a 48 megapixel camera, is video capable with 34 minutes flight time on battery, and can travel at a maximum speed of 20 m/s [3]. Figure A.3 shows the Mavic UAV. The COTS drones are difficult to detect, both from a human as well as an electronic sensor perspective, because they can be made from nonreflective material and can have a small surface area too. Hence there has been considerable work done to develop detection capabilities for counter UAV purposes, not only with mmWave radar but with lidar, acoustic sensors, and combinations of these [4]. Given that UAVs tend to have propellers there has been investigations of their detection using micro-Doppler signatures with radar [5].

There is also a class of UAVs known as suicide drones, which are designed to wait in the vicinity of a target and then attack it once it has been detected. These threats fall into the category of loitering munitions, where the latter can include more sophisticated detection and tracking capabilities. Figure A.4 provides an example of an Israeli-built suicide drone [6], which has a gimballed electro-optic and infrared camera, which is designed to facilitate the

Figure A.2 The Northrop Grumman MQ-4C Triton UAV, used for long-range surveillance, and currently being introduced in Australia for active service. (Source: Royal Australian Air Force.)

Figure A.3 DJI's Mavic 2 Quadcopter, which is a COTS UAV, available for purchase by the general public. (Source: DJI.)

collection of targeting information, and a series of sensors permitting the loitering munition to accurately measure its speed and distance from the target during the final attack.[1]

A.2 Ground-Based Threats

Ground-based systems of interest are known as uncrewed ground vehicles (UGVs) and modern armies have been investigating their utility, since

1. These specifications were kindly provided by Dagan Ari of UVISION.

Figure A.4 Israeli suicide drone. (Source: UVISION.)

they can be used for some dangerous engagements, consequently minimizing human casualties. The Australian Army has been investigating the deployment of UGVs with dismounted combat teams, and as an example Figure A.5 shows a Ghost Robotics UGV supporting Australian Army soldiers during an enemy position clearance operation during an autonomous systems showcase at the Majura Training Area, in Canberra, Australia [7].

UGVs may also be weaponized for fighting purposes, and Figure A.6 provides an example of an Australian-built UGV equipped with sensors and machine guns [8]. This system is semiautonomous with a human-in-the-loop responsible for direct fires engagement.

A.3 Sea-Based Threats

An excellent example of a semiautonomous sea-based system is the U.S. Navy's Sea Hunter, which was initially developed by DARPA [9]. A picture of it is provided in Figure A.7. This is a 135-ton uncrewed surface vehicle (USV) that is capable of littoral as well as high sea operations. It is reported to have a range of 10,000 nautical miles and can remain at sea up to three months.

Figure A.5 An example of a deployed UGV facilitating dismounted combat operations. (Source: Australian Department of Defence.)

Figure A.6 An Australian UGV equipped for fighting operations. (Source: Australian Department of Defence.)

Figure A.7 The USV Sea Hunter, moored at the Joint Base Pearl Harbor-Hickam, Hawaii. (Source: U.S. Navy.)

Figure A.8 The Double Eagle Mark II UUV being recovered by the HMAS Huon. (Source: Royal Australian Navy.)

It was designed for deep sea surveillance and reconnaissance missions, and is able to integrate into the operations of a conventional fleet.

There are also a large number of submarines known as uncrewed underwater vehicles (UUV), designed for a variety of purposes. For example, the Royal Australian Navy's Huon-class minehunter that is equipped with two electrically powered Saab Double Eagle Mark II UUVs, which are used to assist in mine disposal. Figure A.8 shows this UUV in recovery operations [10].

References

[1] https://www.australiandefence.com.au/defence/unmanned/nano-uas-rolled-out-to-army, 2018.

[2] https://www.airforce.gov.au/technology/aircraft/intelligence-surveillance-and-reconnaissance/mq-4c-triton-unmanned-aircraft, 2023.

[3] http://www.dji.com, 2023.

[4] Svanström, F., C. Englund, and F. Alonso-Fernandez, "Real-Time Drone Detection and Tracking with Visible, Thermal and Acoustic Sensors," in *International Conference on Pattern Recognition*, 2021.

[5] Mazumber, J., and A. B. Raj, "Detection and Classification of UAV using Propeller Doppler Profiles for Counter UAS Systems," in *International Conference on Communication and Electronics Systems*, 2020.

[6] https://www.jpost.com/israel-news/us-army-may-use-soon-israeli-designed-suicide-drones-455572, 2016.

[7] https://www.defenceconnect.com.au/land-amphibious/5873-army-sets-up-rico-office-for-robotic-and-autonomous-systems, 2020.

[8] https://www.lowyinstitute.org/the-interpreter/law-when-putting-autonomous-military-platforms-shopping-list, 2021.

[9] https://www.theverge.com/2016/4/8/11391840/us-navy-autonomous-ship-sea-hunter-christened, 2016.

[10] https://www.navy.gov.au/unmanned-systems/double-eagle-mk-ii, 2023.

APPENDIX

B

Contents

References

Irradiance Function for Adaptive Optics

The idea of adaptive optics has its genesis in astronomy, and was first proposed in [1] but did not become a reality until technological requirements were realized in the 1990s [2]. Recall that atmospheric turbulence perturbs the laser beam's wavefront along its propagation path, which can limit the laser's power incident on a target and consequently result in it not achieving its thermal disruption objective. In order to rectify this the principle of adaptive optics can be applied to improve the performance of an optical system by minimizing the effects of incoming wavefront distortions by the use of a compensating deformable mirror [3]. The key to this is to have an underlying sensor that detects variations in the incoming wavefront, and then have this information relayed to a processor that compensates for the variation by adjusting the mirror in order to preserve the beam quality [4]. As remarked in [5] this is an emerging area within HEL DEW. However it is useful to outline how adaptive optics considerations will impact the HEL irradiance function, which is the purpose of this appendix.

Recall the irradiance function defined in (5.14) which is of the form

$$I(t) = \frac{P_T \mathcal{T}(t)}{2\pi R(t)^2 \left(\alpha_D^2 + \alpha_T^2 + \alpha_J^2\right)}. \tag{B.1}$$

The parameter α_T accounts for turbulence in the beam, and is the primary focus here. The following analysis, based in part upon that in [5], outlines how adaptive optics considerations are used to modify this parameter. The key to this is to show how α_T is related to the so-called Strehl ratio [6].

Refer to the ideal case as the situation where there is no turbulence or jitter, and define the corresponding irradiance function

$$I^*(t) = \frac{P_T \mathcal{T}(t)}{2\pi R(t)^2 \alpha_D^2}. \tag{B.2}$$

The Strehl ratio is defined to be the ratio of (B.1) to (B.2) under the assumption that jitter is negligible. Hence it can be shown that this ratio is

$$S = \frac{\alpha_D^2}{\alpha_D^2 + \alpha_T^2}, \tag{B.3}$$

and by solving (B.3) one arrives at

$$\alpha_T^2 = \frac{\alpha_D^2 (1 - S)}{S}. \tag{B.4}$$

Hence once S is specified this can be applied to (B.4). As discussed in [5] the Maréchal approximation [7] can be used under some conditions to estimate the Strehl ratio. In particular

$$S = e^{-\sigma^2}, \tag{B.5}$$

where σ^2 models the total residual wavefront variance due to turbulence after adaptive optics have been utilized. This variance consists of three components. The first accounts for sensor limitation due to signal to noise ratio levels. The second term accounts for the fact that the deformable mirror will not be able to detect precisely any aberrations due to limited actuators. The third component in this variance accounts for delays when an aberration is detected and an appropriate mirror correction is applied. The approximation in (B.5) is valid only when $\sigma \ll 1$.

Under the assumption that (B.5) is valid, one produces

$$\alpha_T^2 = \alpha_D^2 (e^{\sigma^2} - 1), \tag{B.6}$$

and (B.6) may then be applied to (B.1) to produce the irradiance function

$$I(t) = \frac{P_T \mathcal{T}(t)}{2\pi R(t)^2 \left(\alpha_D^2 e^{\sigma^2} + \alpha_J^2\right)}. \tag{B.7}$$

The results in [5] provide information on (B.5) for telescopic measurements, and the parameterizations are not compatible with those employed for the HEL DEW irradiance functions considered in Chapter 5. Hence in order to provide examples of (B.7) and its application to performance prediction, suppose that σ^2 is modeled statistically. In particular, since (B.7) has been derived under the assumption that this variance is less than unity, suppose that σ^2 is uniformly distributed on the unit interval, and that it is independent of the disruption threshold τ. Hence a Bayesian prior distribution is being applied to eliminate σ^2 from the irradiance and consequent probability of disruption. This may be viewed as a somewhat crude result because it does not capture the true inherent characteristics of σ^2. However it is useful for illustrative purposes.

Applying this, the probability of disruption of a target is given by

$$\begin{aligned} \psi(t) &= \mathbf{P}\left(\int_0^t I(s)ds > \tau\right) \\ &= \int_0^1 \mathbf{P}\left(\tau < \int_0^t I(s)ds \middle| \sigma^2\right) d\sigma^2, \end{aligned} \tag{B.8}$$

where conditional probability has been applied, as well as the fact that the density of σ^2 is unity over the unit interval. With the assumption that τ has an exponential distribution with mean τ_0 it can be shown that (B.8) reduces to

$$\psi(t) = 1 - \int_0^1 e^{-\tau_0 \int_0^t I(s)ds} d\sigma^2. \tag{B.9}$$

As in all the previous examples considered throughout the book assume that a threat travels toward the DEW such that its distance to it is given by $R(t) = v(T-t)$, for $0 \le t \le T$. Then by applying (B.9) to (B.7), and applying the result to (B.8) and performing the integration, it can be shown that

$$\psi(t) = 1 - \int_0^1 e^{-\tau_0^{-1} \frac{P_T t}{2\pi v^2 [T-t]} \left(\alpha_D^2 e^{\sigma^2} + \alpha_J^2\right)^{-1}} d\sigma^2, \tag{B.10}$$

provided $0 \le t \le T$, and where the atmospheric transmittance is assumed negligible ($\mathcal{T}(t) = 1$). Expression (B.10) may be evaluated by Monte Carlo simulation observing that it is the mean of the integrand with respect to a uniform random variable on the unit interval.

Suppose, as has been assumed in a majority of the examples considered throughout, that the jitter is negligible. Then (B.10) reduces to

$$\psi(t) = 1 - \int_0^1 e^{-\tau_0^{-1} \frac{P_T t}{2\pi \nu^2 [T-t]} \alpha_D^{-2} e^{-\sigma^2}} d\sigma^2. \tag{B.11}$$

Set

$$\beta(t) = \frac{P_T t}{2\pi \nu^2 [T-t]} \alpha_D^{-2} \tag{B.12}$$

and change variables in the integral through the transformation $\phi = \sigma^2$ then (B.11) is equivalent to

$$\psi(t) = 1 - \int_0^1 e^{-e^{(\log(\beta(t))-\phi)}} d\phi. \tag{B.13}$$

The expression (B.13) has an interesting feature: For any fixed t its integrand is the cumulative distribution function of a random variable with a Gumbel distribution with location parameter $\log(\beta(t))$ and scale parameter unity. Suppose $X(t)$ is this random variable, then if Y is a random variable uniformly distributed on the unit interval then

$$\psi(t) = 1 - \mathbf{P}(X(t) \leq Y) = \mathbf{P}(X(t) > Y). \tag{B.14}$$

Since $X(t)$ and Y are independent by construction it follows that the probabilities $\psi(t)$ may be generated fairly easily through simulation of the random variables $X(t)$ and Y and Monte Carlo summation of the events where $X(t)$ exceeds Y.

One example is provided to illustrate the results, where it is assumed that $P_T = 10^3$ W, $D = 0.8$m, $M^2 = 1$, and that the threat travels at a speed of 40 m/s, taking 120 seconds to reach its intended target (the DEW) if not disrupted. The three wavelength cases used previously are considered as documented in Table 5.1 and the disruption threshold parameter is $\tau_0 = 10^{10}$ W/m^2. The number of Monte Carlo runs for each t used was 10^5. Figure B.1 plots the probability (B.13) for the three wavelength cases. One can conclude that as the wavelength increases it will take considerably longer to defeat the

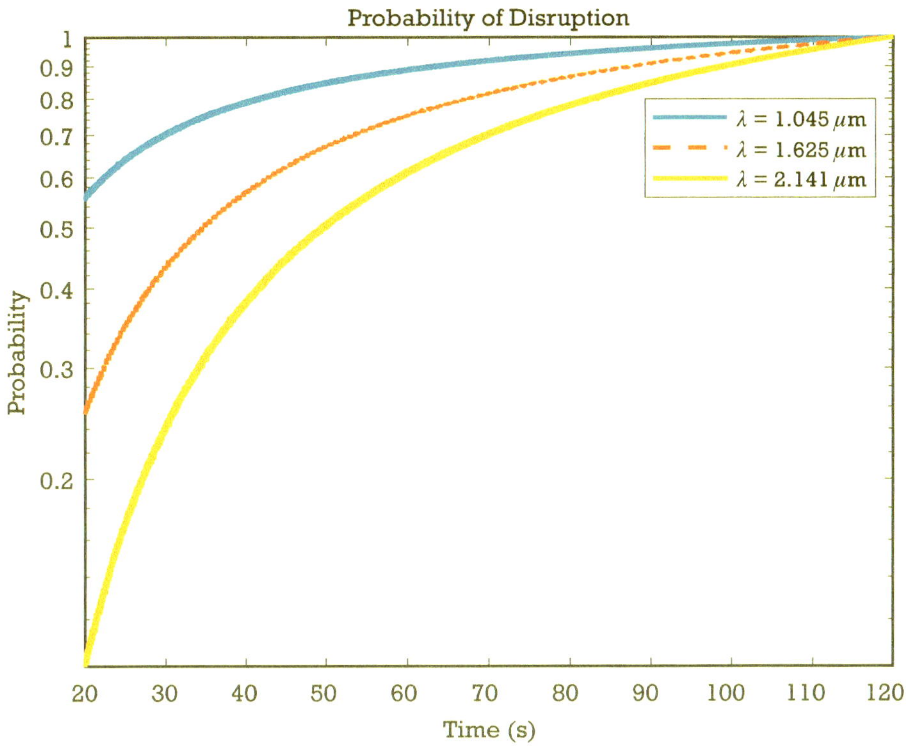

Figure B.1 Plots of the probability of disruption (B.14) for the three wavelengths.

threat. When the threshold is reduced, the probability of defeat increases significantly.

References

[1] Babcock, H. W., "The Possibility of Compensating Astronomical Seeing," *Publications of the Astronomical Society of the Pacific*, Vol. 65, 1953.

[2] Tyson, R. K., *Principles of Adaptive Optics*, Cambridge, MA: Academic Press, 1998.

[3] Nemoto, K., et al., "Transformation of a Laser Beam Intensity Profile by a Deformable Mirror," *Optics Letters*, Vol. 21, No. 3, 1996, pp. 168–170.

[4] Roddier, F., "Curvature Sensing and Compensation: A New Concept in Adaptive Optics," *Journal of the Optical Society of America*, Vol. 27, No. 7, 1988, pp. 1223–1225.

[5] Puent, D., "Integration of Adaptive Optics into High Energy Laser Modeling and Simulation," Technical report, Naval Postgraduate School, Monterey, California, 2017.

[6] Mahajan, V. N., "Strehl Ratio for Primary Aberrations in Terms of their Aberration Variance," *Journal of the Optical Society of America*, Vol. 73, No. 6, 1983, pp. 860–861.

[7] Ross, T. S., "Limitations and Applicability of the Maréchal Approximation," *Journal of the Optical Society of America*, Vol. 48, No. 10, 2009, pp. 1812–1818.

APPENDIX

C

Contents

Reference

Distribution of Delays

Suppose that a series of threats are being processed by a HEL DEW through a G/G/1 queue without reneging. Hence the arrivals are through a renewal process $\{S_n, n \in \mathbb{N}\}$ where each $S_n = X_1 + X_2 + \cdots + X_n$ and every X_j is an independent and identically distributed random variable. Suppose that the server service times are given by the process $\{Y_n, n \in \mathbb{N}\}$ and the series of corresponding delays in the queue customers experience is $\{D_n, n \geq 1\}$ where $D_1 = 0$ since the first arrival is served immediately. Suppose that the threats are on a direct course to the HEL DEW and that threat n travels at a speed of v_n m/s, and will take T_n seconds to reach the HEL if not intercepted. Hence $R_n(t) = \max\{0, v_n(T_n - t)\}$ is the distance between the nth threat and the HEL DEW.

Recall that threats waiting for service by the HEL are still on an intercept course with the HEL DEW. Consequently if the delay exceeds the time of arrival the threat has reached the HEL before being targeted by it. Therefore the probability that D_n exceeds T_n is a useful indicator of whether a threat can defeat the DEW. This can be contextualized as follows. Suppose that a combat team has a series of UAVs available and is aware of the presence of a HEL DEW operating in a given environment. Then it would be useful to ascertain whether the DEW can be overwhelmed with a swarm of UAVs, since one of these may have the capacity to disable the HEL. Hence some rules of thumb would be useful in this situation.

Toward this objective, it can be shown that for the queueing formulation above that the complementary distribution function of the $(n+1)$th delay is given by

$$\mathbf{P}(D_{n+1} \geq t) = \mathbf{P}(\max\{0, U_1, U_1 + U_2, \ldots, U_1 + U_2 + \cdots + U_n\} \geq t), \quad \text{(C.1)}$$

where $U_j = Y_j - X_{j+1}$. This result is derived in [1].

Under certain model parameterizations one may then use (C.1) to gain an understanding of the utility of the DEW to process threats in a time-sufficient manner. An example is warranted to illustrate this procedure. Towards this objective suppose that a HEL DEW is operating as in the queueing example in Section 6.3, except only five threats are present. The arrival rate is still set to unity. The probability (C.1) is evaluated using Monte Carlo simulation with 10^5 runs. The disruption threshold has been set to $\tau_0 = 10^9$ assuming sufficiently shielded threats.

Figure C.1 plots (C.1) for the case where $\nu_n = 40$ and $T_n = 120$ for all n. Hence each curve represents the delay each threat will experience when arriving into the queue. From this figure it is clear that the complementary distribution functions converge to zero for all five threats, which implies that $\mathbf{P}(D_n < T_n)$ is almost certain. This means that under the threat arrival configuration the HEL DEW will defeat all the threats before they have a chance to arrive at the DEW's location.

Suppose that now the first threat has the same arrival process parameters ($\nu_1 = 40$ and $T_1 = 120$) but all other threats travel at a speed of $\nu_n = 100$ m/s, and that they take $T_n = 10$ seconds to arrive at the DEW's location if not intercepted. Figure C.2 provides a plot of (C.1) in this case. As can be observed, the first threat will be defeated but it is almost certain that none of the other threats will be defeated before they arrive at the HEL DEW ($\mathbf{P}(D_n \geq T_n) = 0.9997$ for the remaining 4 threats).

This example is illustrating the way in which a HEL DEW may be defeated in practice. The key to this is to launch the initial threat, knowing it will be served immediately, and then follow it with a series of rapidly traveling threats toward the HEL DEW. Based upon this example, the HEL could be

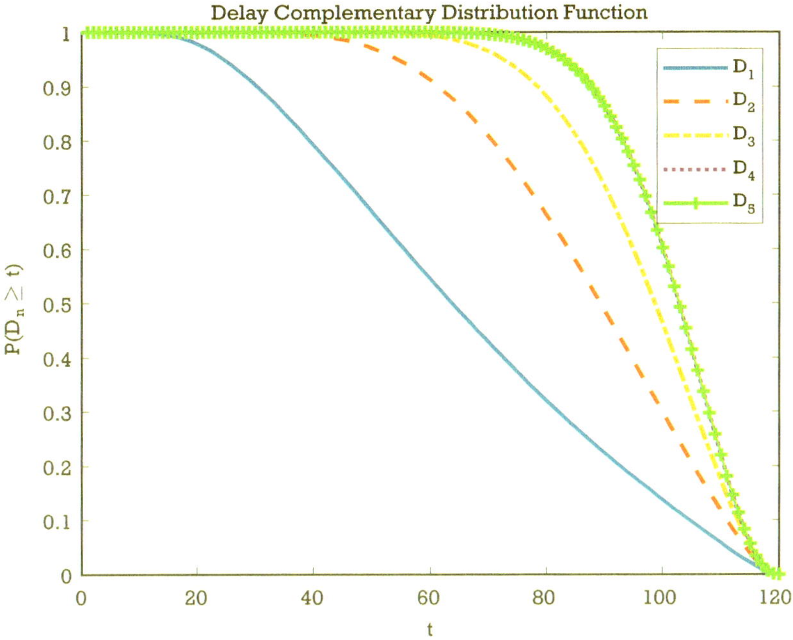

Figure C.1 First example of delay complementary distribution functions.

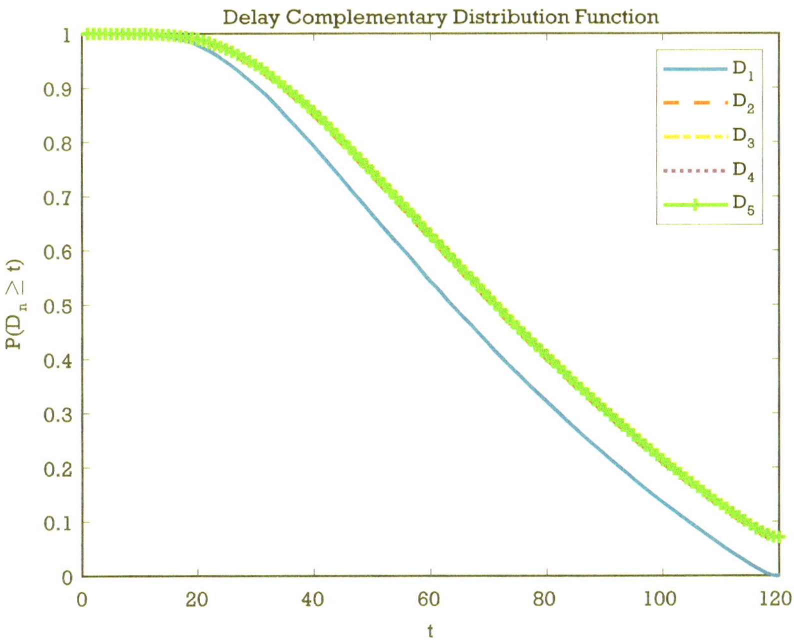

Figure C.2 Second example of delay complementary distribution functions.

defeated by the second threat launched if it traveled under the specified conditions.

Reference

[1] Ross, S., *Stochastic Processes, 2nd Edition*, New York: John Wiley and Sons, 1996.

List of Acronyms

ADS	Active denial system
AFRL	(U.S.) Airforce Research Laboratories
AFV	Armored fighting vehicle
APS	Active protection system
CAPS	Collaborative APS
CFRC	Carbon fiber reinforced composites
CHAMP	Counter Electronics High Powered Advanced Missile Project
COTS	Commercial off-the-shelf
DARPA	(U.S.) Defense Advanced Research Projects Agency
DE	Directed energy
DEW	Directed energy weapon
DSTG	(Australian) Defence Science and Technology Group
EMI	Electromagnetic interference
EMP	Electromagnetic pulse
HEL	High energy laser
HPEM	High-power electromagnetics
HPRF	High-power radio frequency
IEMI	Intentional electromagnetic interference
IID	Independent and identically distributed
IR	Infrared
ISR	Intelligence, surveillance, and reconnaissance
LASER	Light amplification by stimulated emission of radiation
LRAD	Long-range acoustic device
LWS	Laser weapon system
MASER	Microwave amplification through the stimulated emission of radiation

NB	Narrowband
PSD	Power spectral density
RCS	Radar cross section
RF	Radio frequency
RPG	Rocket propelled grenade
SWaP	Size, weight, and power
TEM	Transverse electromagnetic mode
THOR	Tactical high power operational responder
UAV	Uncrewed (archaic unmanned) aerial vehicle
UGV	Uncrewed ground vehicle
U.S.	United States (of America)
USSR	Union of Soviet Socialist Republics
USV	Uncrewed surface vehicle
UUV	Uncrewed undersea vehicle
UWB	Ultrawideband

List of Symbols

c	Speed of light in vacuo 299,792,458 meters per second
$\mathbf{N}$	Natural numbers $\{1, 2, \ldots\}$
$\mathbf{R}$	Real numbers
$\mathbf{R}^+$	Nonnegative real numbers (including zero)
$\mathbf{P}$	Probability
$\mathbf{E}$	Expectation
Var	Variance
$F_X(t)$	Distribution function of random variable X: $F_X(t) = \mathbf{P}(X \leq t)$
$f_X(t)$	Density function of random variable X
$\stackrel{d}{=}$	Equality in distribution
$\mathcal{F}(s)$	Fourier transform
$\mathcal{R}(s)$	Autocorrelation function
$\mathcal{S}(s)$	Power spectral density
$N(t)$	Counting process
S_n	Renewal process
D_n	Queue delay
Y_n	Queue service time
Z_n	Threat reneging time
$M(t)$	Number of threats passed through queueing process by time t
$\mathbf{I}(x)$	Indicator function: takes unity if condition x is true and zero otherwise
λ	Wavelength
$I(t)$	Power density (HPRF) and irradiance function (HEL)
P_T	Peak power
G_A	Antenna gain

A_T	Antenna effective area
$R(t)$	Distance between the effector and target at time t
r_0	Transverse coherence length
c_N^2	Refractive index structure constant
σ_R^2	Rytov variance
α_D	Diffraction pattern parameter
α_T	Atmospheric turbulence parameter
α_J	Jitter parameter
N_B^2	Thermal blooming distortion parameter
$\mathcal{T}$	Atmospheric transmittance function
M^2	Laser beam quality parameter
D	Laser transmitter aperture diameter
R_o	Laser spot size on the target
$w(z)$	Laser beam radius at transverse distance z from origin
z_R	Rayleigh range
n_0	Index of refraction
τ_0	Laser disruption threshold mean parameter

List of Units

dB	Decibel
Hz	Hertz
W	Watt
g	Gram
K	Kelvin
k	Kilo 10^3
M	Mega 10^6
G	Giga 10^9
T	Tera 10^{12}
P	Peta 10^{15}
c	Centi 10^{-2} (when applied as a prefix)
m	Milli 10^{-3} (when used as a prefix, with exception of speed); meter (when applied as a suffix)
μ	Micron 10^{-6}
s	Second
m/s	Meters per second
V/m	Volts per meter
W/m^2	Watts per square meter
W/Hz	Watts per Hz

About the Author

Graham V. Weinberg is a mathematician specializing in probability and stochastic processes, who completed a PhD degree at the University of Melbourne in 2001. He has been employed by the Australian Defence Science and Technology (DST) Group since 2002, and has worked across a number of different domains. In a nutshell, his work at DST Group has been to lead teams investigating the mathematical modeling of defense-related concepts. Most of his career was spent in airborne maritime surveillance radar but over the last five years has shifted to land-based combat modeling and directed energy weapons performance prediction.

He is an associate editor of *IET Electronics Letters* and Elsevier's *Digital Signal Processing*. His first book was published in 2017 (*Radar Detection Theory of Sliding Window Processes*, CRC Press), documenting his extensive research in the area of noncoherent constant false alarm rate radar detection.

Index

A

Acoustic systems, performance prediction, 166
Acronyms list, 187–88
Active Denial Systems (ADS), 32–33
Active protection systems (APS), 165
Adaptive optics
  beam distortion and, 111
  future research, 165–66
  integration, 111–12
  irradiance function for, 177–81
  turbulence and, 131, 165
Advanced Research Project Agency (ARPA), 24
Airborne threats, 169–71
Air Force Research Lab (AFRL) Tactical High-Power Operational Responder (THOR), 30–31, 32
Armored fighting vehicles (AFVs), 25–26
Arrival rates, 99, 100, 104–6
Atmospheric transmittance function, 119, 179
Autocorrelation function, 45
Autovariance function, 45, 57

B

Back door coupling, 65
Beam jitter, 119
Beam quality factor, 119
Beam radius, 116
Beam spreading factor, 119
Bernoulli trials, 93
Boeing YAL-1 airborne laser testbed, 28

C

Carbon fiber reinforced composites (CFRC), 114
Chemical lasers, 111
Complementary distribution function, 74, 153, 155
Conditional probability, 44–45, 97, 179
Continuous probability functions, 40–41
Continuous wave (CW) lasers, 111
COTS UAVs, 169–70
Counter Electronics High Powered Advanced Missile Project (CHAMP), 30
Coupling, 64–65, 81–82
Cumulative distribution function, 40–41, 42–43, 49, 180

D

Damped sinusoidal signal
  about, 76
  absolute value of Fourier transform, 77

effects prediction perspective, 79
frequency bins, 78–79
illustrated, 77
power distribution, 77–78
power spectral density (PSD), 79
pulse train, 76–77
relative signal distribution, 80
See also HPRF DEWs
Delay complementary distribution functions, 185
Delays, distribution of, 183–86
DE systems
about, 17–18
Active Denial Systems (ADS), 32–33
Boeing YAL-1, 28
capability, 12
Epirus Leonidas, 29–30
historical sketch, 20
Laser Weapon System (LWS), 26–28
Long-Range Acoustic Device (LRAD), 33
performance, statistical modeling of, 34–35
in questionable circumstances, 33–34
selection of, 25
Silent Guardian Protection Systems, 32–33
Stryker-mounted laser, 25–26
Tactical High-Power Operational Responder (THOR), 30–31, 32
DE weapons (DEWs)
about, 11, 17
interaction with target, 19
introduction to, 17–20
types of, 18
use of, 12
See also HEL DEWs; HPRF DEWs
Diffraction-limited Gaussian beam, 117
Directed energy (DE), 11, 17–20
Discrete probability function, 40
Disruption probability, 146, 179, 181
Disruption threshold, 112–13, 143
Distribution of delays, 183–86
Distributions, 42–43
Double Eagle II UUV, 174, 175

E

Effective area, 75, 97
Effectors
HPRF, 72
narrowband and wideband performance modeling, 73–82
number to achieve minimum performance, 158–61
number to achieve performance level, 160
single, HEL performance prediction, 141–46
single, HEL performance prediction comparison, 159
Einstein, Albert, 23
Electromagnetic energy, propagation of, 52–55
Electromagnetic interference (EMI), 64
Electromagnetic pulse (EMP)
about, 20–21
CHAMP missile and, 30
fields, impact of, 21
resurgence of interest in, 22
Electronic vulnerability levels, 66–67
Emerging threats
about, 169
airborne threats, 169–71
ground-based threats, 171–72
sea-based threats, 172–75
See also Threats
Epirus Leonidas high-power microwave systems, 30
Events, 40
Exponential distribution, 139

F

Fermi, Enrico, 21
Fiber lasers, 112
Fourier analysis, 56–59
Fourier transform, 59
Friis' transmission formula, 71

Future research
active protection systems (APS), 165
adaptive optics, 165–66
HEL DEWs, 164–66
HPRF DEWs, 163–64
performance prediction for acoustic systems, 166
synchronized DEW systems, 164–65
validation of performance models, 166

G

Gamma distribution, 42
Gas lasers, 111, 114
Gaussian beam profiles, 114–18
Gould, Gordon, 23–24
Ground-based threats, 171–72

H

Hard kill, 19
Havana syndrome, 33
HEL DEWs
about, 12, 18
adaptive optics, 165–66
combat scene illustration, 157
future research, 164–66
in GTK Boxer AFV, 26
infrared (IR) band, 51
kill category, 19
for Naval deployment, 26–28
performance, statistical modeling of, 34–35
synchronized systems, 164–65
UAV swarm elimination with, 110
uniformity guarantee, 161
validation of performance models, 166
HEL irradiance function
about, 66, 109–10
for adaptive optics, 177–81
atmospheric transmittance, 119
beam jitter, 119
beam quality factor, 119
as deterministic, 138
diffraction pattern parameter, 119
distortion number, 129, 131
examples, 121–30
formula, 118, 178
Gaussian beam profiles and, 114–18
jitter introduction, 124
laser damage thresholds and, 112
laser overview and, 110–12
laser power increase, 125, 128
in laser power measurement, 114
model parameterizations, 122
no beam jitter, 123
performance prediction and, 144–46
plots, 123–32
Rytov variance with strong turbulence, 126
Rytov variance with weak turbulence, 127
thermal blooming and, 120–21, 144–45
turbulence without jitter, 128
units of power per surface area, 66
See also High energy lasers (HELs)
HEL performance prediction
about, 137
basis for, 110
disruption threshold, 143
expected number of threats defeated, 156
irradiance function and, 144–46
mean service times, 142
models for thresholds, 138–41
multiple sources on single threat, 155–58
multiple targets, 146–55
number of effectors for minimum performance and, 158–61
queueing theory approach, 146–55
single target and single effector, 141–46
wavelength cases, ‘143
See also HEL irradiance function

High-Energy Laser and Integrated Optical-Dazzler with Surveillance (HELIOS), 27–28
High energy lasers (HELs)
about, 22–24
propagation scene, 115
See also HEL DEWs
High power radio frequency (HPRF)
about, 20–22
DE disruption, 72
DE source, 68
effector, 72
EMP and, 20–21
power density function (PDF), 66, 67–73
power incident measurement, 68
study of effects, 20–21
Historical sketch, 20–25
HPRF DEWs
about, 12, 18
applications, 20
damped sinusoidal signal example, 76–80
disruptive effects, 66
distribution of frequencies, 82
effective area, 75, 97
future research directions, 163–64
IEMI and, 22
kill category, 19
main beam coverage area, 86
narrowband and wideband effector performance modeling, 73–82
objective of, 64
overview, 63
performance assessment, 93
performance modeling, 34, 73–82
power density for WB and UWB signals, 80–82
pulsed power application, 64
pulse power components, 22
radio and microwave bands, 51
source power, 73–74
UAV arrival into the beam, 85
UAV defeat near, 101–2
wavelength and field strength, 65
HPRF performance prediction
about, 85–86
conclusions, 107–8
curves, 88
four examples, 87
narrowband case, 86–89
one threat, 86–92
threats arrive linearly, 101–7
threats arrive simultaneously, 92–96
threats arrive through renewal process, 96–101
ultrawideband case, 89–92
vulnerability thresholds and, 95, 96, 101
wideband case, 89–92
See also HPRF DEWs

I

Independent and identically distributed (IID) variables, 41
Infrared (IR) band, 51
Intentional electromagnetic interference (HEMI), 22
Interarrival times, 47, 49
Inverse cumulative distribution, 146
Inverse Fourier transform, 56–57
IR cameras, 113
Irradiance functions
for adaptive optics, 177–81
atmospheric transmittance, 119
beam jitter, 119
beam quality factor, 119
as deterministic, 138
diffraction pattern parameter, 119
distortion number, 129, 131
examples, 121–30
formula, 118, 178
Gaussian beam, 115
integrating over circle, 115–16
jitter introduction, 124
laser power increase, 125

laser power increase to compensate for jitter, 128
in laser power measurement, 114
model parameterizations, 122
no beam jitter, 123
performance prediction and, 144–46
plots, 123–32
Rytov variance with strong turbulence, 126
Rytov variance with weak turbulence, 127
thermal blooming, 120–21
turbulence without jitter, 128
units of power per surface area, 66
See also HEL DEWs; HEL performance prediction
Irradiance levels, selection of material and, 113

J
Jitter, 119, 123–24, 128

L
Lasers
beam angle of divergence, 116
categories, 111
chemical, 111
continuous wave (CW), 111, 112
damage thresholds, 112–14
defined, 110
fiber, 112
gas, 111, 114
Gaussian beam profiles, 114–18
overview, 110–12
power measurement, 114
pulsed, 111
quantum cascade, 113
solid-state, 111
wavelength choice, 112
See also High energy lasers (HELs)
Laser spot sizes, 116
Laser Weapon System (LWS), 26–28
Logarithmic functions, 159–60
Long-range acoustic devices (LRAD), 33, 166

M
Maiman, Theodore H., 24
Maréchal approximation, 178
Mavic UAV, 170, 171
Maxwell's equations, 50–60
Mean arrival rate, 106
Mean number defeated (threats), 107, 149
Mean service times, 140–41, 142
Micro-UAVs, 169
Microwave bands, 51
M/M/1 model, 49
Moments of the random variable, 41
Monte Carlo simulation, 180
MQ-4C Triton UAV, 169, 171
Multiple targets (HEL performance prediction), 146–55
Multiple threats (HPRF performance prediction)
about, 92
arrive linearly, 101–7
arrive simultaneously, 92–96
arrive through renewal process, 96–101
converging onto HPRF DEW, illustrated, 94
defeated in given time period, 102–3
vulnerability thresholds, 95, 96
See also HPRF performance prediction

N
Narrowband (NB) signals, 75
Narrowband case (HPRF performance prediction), 86–89
Nulls, 53

O
One threat (HPRF performance prediction)

about, 89
four examples, 87
narrowband case, 86–89
ultrawideband case, 89–92
wideband case, 89–92
See also HPRF performance prediction

P

Parameterizations, 122, 151, 179, 184
Peak power density, 68–69, 70
Performance prediction
for acoustic systems, 166
HEL, 137
HPRF, 85–86
Physics
introduction to, 50–60
propagation of electromagnetic energy and, 52–55
signals and Fourier analysis and, 55
wavelength and frequency, 51–52
Point probabilities, 49, 98
Poisson model, 43
Poisson process, 46, 47, 104, 148
Polystyrene, 114
Power density, for wideband and ultrawideband signals, 80–82
Power density function (PDF)
distribution function, 71–72
HPRF, 66, 67–73
HPRF on threat over observational window, 92
plots, 73
Power spectral density (PSD)
about, 57
damped sinusoidal signal, 79
determination, 81
example, 60
Probability
basics, 40–41
conditional, 44–45, 97, 179
density, 41
of disruption, 146, 179, 181
distributions and properties and, 44
evaluation, 44
point, 49
reneging, 105
statistical, 44–45
Probability density function, 42
Probability mass function, 43
Propagation, electromagnetic, 52–55
Properties, 42–43
Pulsed lasers, 111
Purpose and scope, this book, 34
Pythagorean theorem, 157

Q

Quantum cascade laser, 113
Queueing theory
about, 47
approach formulation, 146
assignment algorithm, 149
complementary distribution function, 153, 155
in HEL performance prediction, 146–55
illustrated, 48
mean numb of threats defeated and, 149
mean of individual queue, 154
parameterizations and, 151
processing multiple threats through, 148
rate of arrivals and, 147
service times of threats and, 152–53
single-server model, 48
threat reneging times and, 147
threshold parameter, 153
See also Stochastic processes

R

Radar cross section (RCS), 71
Radio frequency bands, 51
Random variables
about, 41
determining probabilities associated with, 47
gamma distribution, 42

IID exponential, 42
independent, 97
moments of, 41
service time to target, 103
Rayleigh range, 116
Reneging probability, 105
Renewal process
about, 47
arrivals through, 96–101
theory, 98
RF energy, 52
Rocket-propelled grenades (RPGs), 101
Rytov variance, 120, 126

S

Sampling rate, 81
Sea-based threats, 172–75
Service time to target, 103
Signals and Fourier analysis, 57
Silent Guardian Protection Systems, 32–33
Simpson's rule, 78
Simulation, 39, 180
Size, weight, and power (SWaP), 26
Soft kill, 19
Solid-state lasers, 111
Statistical probability, 44–45
Stochastic processes
autocorrelation function, 45
autovariance function, 45
defined, 45
fundamental, 45–47
mean measure, 45
Poisson process, 46, 47, 104, 148
queueing theory, 47–50
renewal process, 47
states of, 45
stationary, 45–46
theory, 39
wide-sense stationary, 46
Strehl ratio, 178
Suicide drones, 170–71, 172
Symbols list, 189–90
Synchronized DEW systems, 164–65

T

Tactical High-Power Operational Responder (THOR), 30–31, 32
Targets
disruption threshold, 112–13
multiple (HEL performance prediction), 146–55
probability of disruption, 179, 181
vulnerability parameter, 100
Technical Research Group (TRG), 24
Thermal blooming
about, 120
approximate distortion parameter, 120–21
effects, 118
formula, 120
HEL performance prediction and, 144–45
Threats
arrival rates, 99, 100, 104–6
arrive in effective area, 97
arrive linearly, 101–6
arrive randomly, 96
arrive simultaneously, 92–96
arrive through renewal process, 96–101
converging onto HPRF DEW, illustrated, 94
coupling cross section parameter, 96
defeated in given time period, 102–3
emerging, 169–75
expected number defeated, 156
ground-based, 171–72
lifetime, arrival rate and, 104, 105, 106
linear trajectory, 99
mean number defeated, 107, 149
mean service times, 140–41, 142
multiple (HPRF performance prediction), 92–107
number defeated, 98
one (HPRF performance prediction), 86–92

probability of defeating, 93
sea-based, 172–75
service times, 103, 152–53
single, multiple sources (HEL performance prediction), 155–58
in straight trajectory, 102, 103
waiting for service, 183
3-dB beamwidth, 54
Threshold parameter, 153
Townes, Charles H., 23
Transverse coherence length, 120
Transverse electromagnetic (TEM) modes, 52
Trapezoidal approximation, 78
Turbulence
adaptive optics and, 165
irradiance function and, 178
laser beam wavefront and, 177
Rytov variance with, 126, 127
without jitter, 128

U

Ultrawideband (UWB) signals, 75
Ultrawideband case (HPRF performance prediction), 89–92
Uncrewed aerial vehicles (UAVs)
COTS, 169–70
countering swarms, 65
defeat near HPRF DEWs, 101–2
forensic tracking of, 65
IR cameras, 113
micro, 169
optical sensors, 113
problem posed by, 11–12
suicide drones, 170–71, 172
vulnerabilities, 67
Uncrewed ground vehicles (UGVs), 171–72, 173
Uncrewed surface vehicles (USVs), 172–75
Uncrewed underwater vehicles (UUVs), 175
Units list, 191
U.S. Navy Laser Weapon System (LWS), 26–28
U.S. Navy Sea Hunter, 172, 174

V

Validation of performance models, 166
Volts per meter, 66
Vulnerability thresholds, 67, 95, 96, 101

W

Wavefront curvature, 116
Wavelength and frequency, 51–52
Wideband case (HPRF performance prediction), 89–92

Artech House Electronic Warfare Library

Dr. Joseph R. Guerci, Series Editor

Active Radar Electronic Countermeasures, Edward J. Chrzanowski

Antenna Systems and Electronic Warfare Applications, Richard A. Poisel

Battlespace Technologies: Network-Enabled Information Dominance, Richard S. Deakin

Cognitive Electronic Warfare: An Artificial Intelligence Approach, Karen Zita Haigh and Julia Andrusenko

Developing Digital RF Memories and Transceiver Technologies for Electromagnetic Warfare, Phillip E. Pace

Digital Techniques for Wideband Receivers, Second Edition, James Tsui

Directed Energy System Performance Prediction, Graham V. Weinberg

Electronic Intelligence, Richard G. Wiley

Electronic Intelligence: The Analysis of Radar Signals, Second Edition, Richard G. Wiley

Electronic Warfare for the Digitized Battlefield, Michael R. Frater and Michael Ryan

Electronic Warfare in the Information Age, D. Curtis Schleher

Electronic Warfare Receivers and Receiving Systems, Richard A. Poisel

Electronic Warfare Receiving Systems, Dennis Vaccaro

Electronic Warfare Signal Processing, James Genova

Electronic Warfare Target Location Methods, Richard A. Poisel

ELINT: The Interception and Analysis of Radar Signals, Richard G. Wiley

Emitter Detection and Geolocation, for Electronic Warfare, Nicholas A. O'Donoughue

EW 101: A First Course in Electronic Warfare, David L. Adamy

EW 102: A Second Course in Electronic Warfare, David L. Adamy

EW 103: Tactical Battlefield Communications Electronic Warfare, David L. Adamy

EW 104: EW Against a New Generation of Threats, David L. Adamy

EW 105: Space Electronic Warfare, David L. Adamy

Foundations of Communications Electronic Warfare, Richard A. Poisel

Fundamentals of Electronic Warfare, Sergei A. Vakin, Lev N. Shustov, and Robert H. Dunwell

Information Warfare and Electronic Warfare Systems, Richard A. Poisel

Introduction to Communication Electronic Warfare Systems, Richard A. Poisel

Introduction to Electronic Defense Systems, 3rd Edition, Filippo Neri

Introduction to Electronic Warfare, D. Curtis Schleher

Introduction to Electronic Warfare Modeling and Simulation, David L. Adamy

Introduction to Modern EW Systems, 2nd Edition, Andrea De Martino

Military Communications in the Future Battlefield, Marko Suojanen

Modern Communications Jamming Principles and Techniques, Richard A. Poisel

Modern Communications Receiver Design and Technology, Cornell Drentea

Practical ESM Analysis, Sue Robertson

Radar Electronic Countermeasures, Stephen L. Johnston

RF Electronics for Electronic Warfare, Richard A. Poisel

Synthetic Aperture Radar and Electronic Warfare, Gisela Goj

Tactical Communications for the Digitized Battlefield, Michael Ryan and Michael R. Frater

Target Acquisition in Communication Electronic Warfare Systems, Richard A. Poisel

For further information on these and other Artech House titles, including previously considered out-of-print books now available through our In-Print-Forever® (IPF®) program, contact:

Artech House
685 Canton Street
Norwood, MA 02062
Phone: 781-769-9750
Fax: 781-769-6334
e-mail: artech@artechhouse.com

Artech House
16 Sussex Street
London SW1V 4RW UK
Phone: +44 (0)20-7596-8750
Fax: +44 (0)20-7630-0166
e-mail: artech-uk@artechhouse.com

Find us on the World Wide Web at: www.artechhouse.com